THE ODDS
On Virtually Everything

THE ODDS

On Virtually Everything

Compiled by the editors of
Heron House

Preface by Richard Scammon

Commentaries by Peter J. Verstappen

G. P. Putnam's Sons
New York

Library of Congress Cataloging in Publication Data

The odds on virtually everything.

    1. Probabilities—Popular works.   I. Scammon, Richard M.
QA273.15.032   031'.02   79-20218
ISBN: 0-399-12483-7

ACKNOWLEDGEMENTS

No enterprise of this size could be completed without the active assistance of hundreds of people and their organizations. To list them all would take up much of the existing book. Sadly, we can't do so but we are most grateful to one and all. Virtually without exception those contacted were helpful in the extreme while many went out of their way to put us in touch with others who might (and normally did) help.

Special thanks are due to the following:

Our patient (against all odds) editor Sam Mitnick. The book was his idea in the first place—he steadfastly refused to disavow the feasibility of the project in the face of severe provocations. His many contributions including bringing my writing style closer to the prevailing patterns of English prose have been valuable. Dick Scammon of the Elections Research Center whose warmth and good humor match his unparalleled knowledge; Pat Kennedy of the Civil Aeronautics Board who contributed much sound advice and some invaluable contacts; Len Wood of the Gallup Organization, a perceptive, patient critic and a good friend; Ed Jones at the Department of Justice who went out of his way to come up with the goods; Hank Bernhard of Ogilvy and Mather who, as always, dropped everything to set matters right; Cathy O'Brien of the Population Division, Bureau of the Census and Jack Recht and Wayne Gregory of the National Safety Council, who, along with many of their colleagues, were exceptionally generous at sharing and clarifying figures.

Others whose contribution we'd especially like to acknowledge are: Jim Boyle, U.S. Secret Service—John Bousquet, U.S. Trust Company—Mike Carberry, Henry J. Kaufman and Associates—Mike Cushing, National Transportation Safety Board—Clarence Ditlow, Center for Automotive Safety—Connie Dresser, Department of Health & Human Resources—Angela Fairchild, National Clearing House for Drug Abuse—W. Vance Grant, National Center for Education Statistics—Paul Gilliland, National Auto Theft Bureau—Aline Herman, Department of Labor—Ben Kelly, Insurance Institute for Highway Safety—Glenn Phillips—Educational Testing Service—Nancy Thompson, National Organization of Women—and—Johnnie Walker, Los Angeles Times Syndicate.

Additional thanks to Find/S.V.P., whose research was of great use in preparing half of this text.

More than honorable mention must go to the gang at Herb Field Studios who set the type and prepared this book for printing.

In our own organization Kathi Paton performed yeoman service as editor. Jeri Calviello, Pam Mayle, Peggy Martinian, Amy Verstappen, Marie Tamplenizza and Rhoda Erath all assisted with odds and ends.

In the final analysis books, like children, need parents to hover over, encourage and love them. This book would not have seen the light of day without the remarkable talents of one such person. Deb Loveless has coordinated and cajoled, designed and typed, edited and improved with unfailing good humor and great verve. She is the epitome of an old saying which in essence goes—"If you need to get something done well, ask a busy person."

Peter Verstappen

TABLE OF CONTENTS

PREFACE

By Richard M. Scammon, Director of the Census, 1961-1965

Some fifty years ago Agatha Christie wrote one of her mystery novels, including in it a character who called everything in the world "subtle." Subtle *The Odds* book is not. Direct, sharp, frontal, numerate...all these words may apply to this Heron House work, but certainly not subtle.

What the editors have put together for the reader is a series of well-founded ratios, ratios which tell us *what are the odds* on this or that...on your owning a yacht, on suffering a heart attack, on being elected to Congress, on just about anything, anywhere, anytime.

A lot of us are betting people, not perhaps chronic race track betters or bingo bean-pushers or even weekly poker players, but bettors nonetheless. "I'll give you two-to-one that my old school beats yours this fall in the Thanksgiving game" or "Your candidate hasn't got a chance and here is ten that says he won't carry ten states in November." We've all heard this kind of talk, and many of us have been the winning—or losing—participant in one of these bets.

But *The Odds* is a lot more than just a betting book. A lot of the things you'll find in *The Odds* don't even deal with the things on which people usually bet. It would be interesting to know how many of the huge number of things covered by *The Odds* have actually been the subject of a money bet known to our friend Jimmy the Greek, but there must be many on which no one ever put down a dime, thin or otherwise.

There must also be many things on which not even protective insurance has ever been taken. Most of our readers have seen press stories about the actress who insures her legs or the opera tenor who insures

i

his voice. Even though the stories may be largely a matter of publicity, the insurance need may be very real. For an outdoor production of *Aida*, as one example, insurance against rain is much more than publicity. It is a bet on the odds of having to postpone or cancel a costly performance and refund the ticket money of a number of the intended audience.

Many of the odds quoted in this volume are on pretty serious matters —marriage, birth and death, jobs, and the like. Some are "fun" odds: on cooking your own goose, on finding a pig in a poke, on being in stitches. But serious or fun, the figures are backed up by authority, and our readers are invited to get in touch with these people if they need any further refinement or detail on what *The Odds* book tells them in general.

Naturally, some of these odds are changing even as *The Odds* goes to press. Nothing remains static forever, and the proportions of us who do this or that, or who are that or this, are always in flux. This year of publication—1980—will see two big events which may change more than a few of these odds. In April the 1980 Census counts us all, and in November we go through our quadrennial election.

In this last will be tested one of the strangest odds of all: the chances of a President being elected in a year ending in zero dying while in office. Since 1840 the odds are total: every man chosen in such a year—1840, 1860, 1880, 1900, 1920, 1940, 1960—has died while still President of the United States. We'll see what happens this fall, and in the years which follow, to this 140-year old odds figure.

A final word: we can't settle odds arguments for you. We believe our figures are as sound as good research can make them. But don't bet on it!

INTRODUCTION

By Peter J. Verstappen

Any book dealing with odds owes it to the reader to start with an explanation of the concept. After all, the odds are you find odds confusing. Put simply, odds are the exact number we attach to a future possibility to state how likely it is to occur. If you shop every other day, the odds of your shopping on any given day are 50/50: in other words on a fifty percent chance the odds are even. If you shop only on every third day, we can find the odds of your shopping on any given day as the relationship between two days of non-shopping and one day of shopping. Thus, the odds of your going shopping on any given day are 2-1 against.

The above examples demonstrate simply all the elements needed to compute the odds. First, you need to know the number of times an event actually occurs compared to the number of times it does not occur (in this case shopping vs. not-shopping), and you need a constant time frame for the two events (in this case 3 days).

Let's take the game of football as another example. Here the game is the time frame. The past performance of each of the two teams can provide us with the basic data we need to predict the odds on one team's performance versus that of its opponent. Thus, if team A had won its last nine games and team B had lost nine games the odds would heavily favor team A. The precise odds would depend on the relative records of the two teams. Suppose that team A had won nine of its last 10 games while B had lost nine of its last 10 games. The initial calculation of odds would then be 9-1 x 9-1 = 81-1 that team A would win based on past performance alone.

But oddsmakers take into account other factors which may have an effect on the outcome. Returning to our football game, suppose that even though team A had a vastly superior record to B's, it lost every game it had ever played against a team with a left-handed quarterback, and also assume that team B has one. This new factor would lower team A's chances of winning. Let us further suppose that for years team B had won all of their final home games of the season and this contest would be their final home game. Again this would reduce team A's chances and increase the odds in favor of team B winning.

Combining all these factors which an odds setter believes to have an effect on the outcome of an event, and giving each factor a number in relation to the other factors is the essence of predicting odds.

This brings up the next principle of odds. For any given odd in favor of something happening there is a reverse odd in favor of it not happening. If we are discussing dogs and it is known that one out of 100 dogs bite, this can be discussed in one of two ways:

A. The odds that a dog won't bite   99-1 for
B. The odds that a dog bites   99-1 against

To determine the odds contained in this book, we embarked on a two year detective hunt to obtain the most accurate facts about various data and then made every effort to compute them in the proper way.

Because statistics affect decisions (and more often than now how money is spent), various interest groups will often use different numbers or the same numbers in a different way. Is this dishonest? Not necessarily. Often it simply reflects an honest difference of opinion. The 1980 Census provides a perfect example of one such honest difference. California, New York and Florida have large illegal immigrant populations. Each of these states must provide public services for these groups. The allocation of federal money back to each state is based on the census count of legal citizens. All three of these states argue that the immigrants should be included in the count. It's in their financial interest to do so. Other states argue that these people are not part of the American population and thus should not be included. Their argument is designed to prevent funds earmarked for programs in their states being diverted elsewhere. Statistically there are "two" population counts. As you can imagine if a basic question like "what is the population?" is subject to challenge on legitimate grounds, other population samples can be even more readily challenged. For this reason we have carefully sourced all of the odds cited in this book. Where conflicting data has come to the editors' attention we have analyzed the alternatives, sought expert advice and made a choice. Almost without exception we have chosen U.S. government figures in preference to others. Having said that, it's distressing—but true—to state that sometimes different divisions of the federal government produce widely varying statistics on the same topic.

The reasons for these differences range from simple mathematical errors (on several occasions civil servants pleaded with us not to reveal a goof) to highly sophisticated projections of samples.

In this latter case, you take a sample of the population and expand it proportionately to national size. Depending on both the sample size and the research techniques employed, there is the possibility of error creeping in. Researchers express this concept as "the data is reliable to a plus or minus tolerance." We consciously disregarded the plus or minus factor in an effort to make this a book for everyone, not just the specialized researcher. By the same token on long odds of 100 and over we did not carry our figures out to a decimal because this seemed too academic.

Computing the odds has presented us with some thorny problems. None has provided greater controversy among the editors than the question of what base group to predict the odds against. The F.B.I. reports the number of rapes based on a figure per thousand of total population; we altered their number to a figure per thousand women to compute the odds. A more sophisticated alternative was based on the fact that experts estimate that for every reported rape a second takes place, so we expanded the figure proportionately to arrive at a national figure.

Often what seems to be the likely base group is not comprehensive enough: you wouldn't think the odds of being a murderer should include children under the age of ten. Yet children under this age committed 75 murders in the year covered. Thus, as in our murderous example, what may not be expected as our base group is in fact a logical use of the data we have gathered.

In getting the facts needed to determine the odds we have been overwhelmed by the generosity of those in business, education, government and the research field and their willingness to help us. Almost invariably when we queried someone and explained what we were up to the response was "that sounds like fun—what can I do to help?". But government seems to be particularly unfashionable at the moment in part because bureaucrats are seen as unresponsive to the public's needs. We found nothing to be farther from the truth. Many of Nader's Raiders are now in responsible government positions, often in the very agencies they monitored before. They've not forgotten the public's need to know.

In compiling the present work we had to calculate several hundred thousand statistics. We did this twice and the results have been doublechecked by outside experts. We'd like to think they're error free but the odds are against it. The editors of Heron House are solely responsible and apologize in advance for any inadvertant goofs.

Finally, if preparing The Odds has taught us anything, it is how very philosophic odds are and the vital role they play in everyday life. Predicting the price of gold and having children, tomorrow's defense budget and the weekend's weather are all predicated on pegging the odds. There are no odds to be set on whether or not you'll enjoy this book, but we had a terrific time working them out, and wondering at the results. It is our hope that the odds favor your enjoying them too.

THE ODDS
On Virtually Everything

FOR THE LOVE OF MONEY
Success, Wealth and Taxes

Thinking of hitting the jackpot on that 25-cent slot machine? You'd better have more than a pocketful of quarters. On the average only one crops up for every $222.25 played. The odds on hitting the jackpot in life are far better. One out of 424 Americans is a millionaire. This varies widely by state as you'll see. *For the Love of Money* lets you compare your good fortune with your fellow man's. It also provides practical information, such as the chances in poker of filling an inside straight (slim) and of having your tax return audited (far slimmer).

Do you sometimes buy an extra box of detergent to increase your odds of winning a $50,000 sweepstakes? It does help your odds, but they're still stacked against you: over six and a half million to one. Broadway's enjoying its best season in years, with people standing in line to become backers, but the odds of getting your money back are over twice as long as those who follow the roller coaster ride of commodities fortunes.

We offer some odds that you can manipulate. For example, if you pay the I.R.S. with a check that bounces your chances of joining the other 91,000 people who faced prosecution for this offense are exceptionally good. But if money makes the world go round, checks make it bounce in its orbit: over 214 million duds are cashed every year.

Finally, we look at every man's dream, printing your own money. For those who try it it turns out to be a nightmare with high odds on being caught and lousy odds on making decent money before you're nailed.

WHAT ARE THE ODDS ON COINING MONEY?

The U.S. Treasury Department informed us that currently 2,344 men and women are employed by the U.S. Mint. We compared them to the estimated current total work force.

The Odds: **44,794-1** against

Source: Bureau of Labor Statistics, *Employment and Earnings,* Monthly.

WHAT ARE THE ODDS THAT A FOOL AND HIS MONEY ARE SOON PARTED?

According to the Las Vegas Visitor's Bureau the average stay in their fair city is 3.5 days, and the almost 12 million tourists leave 3.3 billion dollars behind (about $280 a head).

The Odds: **11,778,111-1** for

Source: Data from Las Vegas Visitor's Bureau.

WHAT ARE THE ODDS ON BECOMING A MILLIONAIRE?

Thirty years ago there were just 13,000 millionaires in the United States; one for every 11,300 citizens. Today there are 520,000 people with a net worth of over one million. The odds are based on 2.36 millionaires per thousand people.

The Odds: **423-1** against

The U.S. Trust Company recently released a study on millionaires in 38 states. Several surprising facts turned up. The first is the rapid growth of the whole category. Between 1978 and 1979 the number of millionaires jumped by a hefty 15%, adding 70,000 to this charmed circle. (It's hardly an exclusive circle anymore since it's almost one quarter of one percent of our total population.)

The boom in land values is the principal underlying cause for the rapid growth of the very rich. Large landowners in our farming and ranching states have, in many cases, doubled and trebled their net worth in recent years.

Thus while our most populated states have the greatest absolute number of millionaires, on a per thousand population basis these states are well down the list. If you're out to bag a millionaire, head for Idaho. The state has 23,797 high rollers in a population of just 893,000. Maine, North Dakota and Montana are next in line. If you abhor wealth, Wyoming's for you. There are 5,262 citizens for every millionaire there compared to almost one millionaire for every 37 citizens in Idaho.

The Odds: against

		Millionaires
Arizona	2,756	858-1
Colorado	6,868	394-1
Connecticut	10,811	290-1
Delaware	1,946	303-1
Idaho	23,797	36.5-1
Illinois	31,138	364-1
Indiana	24,345	218-1
Iowa	11,602	252-1
Kansas	6,822	346-1
Kentucky	4,571	774-1
Maine	8,337	128-1
Maryland	3,220	1,514-1
Massachusetts	18,015	322-1
Michigan	14,029	653-1
Minnesota	22,873	174-1
Missouri	4,864	989-1
Montana	2,518	309-1
Nebraska	10,462	151-1
New Hampshire	603	1,428-1
New Jersey	26,565	276-1
New Mexico	2,825	437-1
New York	51,031	352-1
North Carolina	9,416	594-1
North Dakota	4,598	144-1
Ohio	27,607	387-1
Oklahoma	4,050	713-1
Pennsylvania	15,318	775-1
Rhode Island	1,449	648-1
South Carolina	5,401	542-1
South Dakota	701	1,009-1
Tennessee	11,705	369-1
Texas	21,051	623-1
Utah	665	1,960-1
Vermont	1,105	447-1
Virginia	6,769	774-1
West Virginia	1,046	1,817-1
Wisconsin	19,006	246-1
Wyoming	84	5,262-1

Source: Data from U.S. Trust Company of New York.

WHAT ARE THE ODDS ON MAKING MONEY BY SPECULATING IN COMMODITIES?

For many reasons, the answer is "poor." If you spread your risk over a number of commodities (which brokers advise you to do) you will be eaten alive by commissions as you buy and sell. Consider a $10,000 account which trades 26 times a year at the maximum overall margin of $100,000. At an average of 0.3% commission per "round trip" of buying and selling you'll pay $5,200 in commissions alone. That's 52% of your investment gone.

This and other middleman's fees guarantee that there isn't a winner for every loser. One study showed that of all money put in by losers 36% went to commissions, 23% to the changing house and only 41% to winners.

The Odds: **1.8-1** against

WHAT ARE THE ODDS ON HITTING A JACKPOT?

Have you ever noticed how tantalizingly close you come to hitting the jackpot on a slot machine? They're all planned that way. There are twenty squares on each of the three wheels. Typically the middle one has nine bells out of the 20, meaning that the odds on your middle wheel turning up a bell are almost even. But the outer wheels have only one bell out of 20 alternatives. The calculation looks like this: 9/20 x 1/20 x 1/20.

The Odds: **889-1** against

Source: Data from University of Nevada.

WHAT ARE THE ODDS ON WINNING MONEY IN A CONSUMER SWEEP-STAKES?

The odds are entirely dependent on the number who are exposed to the offer (normally via direct mail or print advertising), the number of entrants who consequently respond, and both the extent of and attraction of the prizes (which obviously affects response). Contacts with a major sweepstakes clearing house, Publishers' Clearing House, several consumer goods companies and the Reader's Digest all confirm the above. The Reader's Digest gave us the odds of winning their current sweepstakes.

The Odds: against

$	5	1,595-1	$ 10,000	6,400,000-1
	10	2,560-1	20,000	12,800,000-1
	100	141,333-1	50,000-	
	5,000	4,266,000-1	118,000	20,000,000-1

Source: The Reader's Digest

WHAT ARE THE ODDS OF APPEARING ON A TV QUIZ SHOW?

According to a number of contestant coordinators, it all depends on the show you want to get on. In our sample Card Sharks is the most difficult (one out of 150 gets on the show), while $20,000 Pyramid is easiest (eight out of 100 make it in front of the camera).

The Odds:		
	Card Sharks	**149.0-1** against
	Family Feud	**60.0-1** against
	Password Plus	**50.0-1** against
	$20,000 Pyramid	**11.5-1** against

Source: Data from TV contestant coordinators.

WHAT ARE THE ODDS OF OWNING YOUR OWN RACEHORSE?

The Horseman's Benevolent and Protective Association lists 45,000 individual racehorse owners for 1978. The odds are based on the estimated 77 million households in the country.

The Odds: **1,710-1** against

Source: Data from Horseman's Benevolent and Protective Association.

IF YOU DO OWN A RACEHORSE WHAT ARE THE ODDS ON YOUR HORSE WINNING THE TRIPLE CROWN?

The odds are based on the 31,326 foals registered in 1978 and the three wins needed.

The Odds: **98,129-1** against

Source: Data from Horseman's Benevolent and Protective Association.

WHAT ARE THE ODDS OF YOUR FAMILY OWNING A YACHT?

As of the first of January 1978, there were 49,183 registered yachts in the United States. Some citizens own more than one yacht while some registered yachts are owned by foreign nationals.

The Odds: **1,511-1** against

Source: Data from the National Association of Engine Boat Manufacturers.

WHAT ARE THE ODDS ON BECOMING A STAR ON BROADWAY?

There are currently 28,000 members of Actors Equity (up 10,000 members in the last five years). According to this organization about 20% of the membership is working in all areas of the field—Broadway, Off-Broadway, Off-Off Broadway, touring companies, etc.—at any given moment. So the odds on acting at any given moment are four to one against.

In total there are currently about 400 performers appearing in some 59 plays. Of these no more than 15 are big enough hits to confer star status on one man and one woman.

The Odds: **932-1** against

Source: Actors Equity Association, New York; League of N.Y. Theaters and Producers, New York.

WHAT ARE THE ODDS OF STARRING ON TV OR IN THE MOVIES?

We took an old-fashioned but practical definition of what constitutes a star: money. Then we asked the Screen Actors Guild (it's impossible to become a star if you're not a member) how many of their 39,000 members earn over $100,000 a year. Only 0.6%, or 234 members, make that much. Incidentally, just 663 members (1.7% of the total) earn over $50,000 a year.

Latest available census figures show that 968,000 people are employed in the category of "Entertainment and Recreational Services." So just 4% have made it into the Screen Actors Guild. By way of comparison 110,000 of them have made it to the unemployment office. At 11.4% this group has the highest unemployment rate of any job category.

So the odds depend on whether you want to compare it with the population at large (we chose those 16 and over), those who are already in this employment category or S.A.G. members.

The Odds: Population 16 and over   **386,752-1** against
            Those in employment
                category   **4,136-1** against
            Screen Actors Guild member   **167-1** against

Source: Screen Actors Guild and Department of Labor data.

WHAT ARE THE ODDS OF HAVING A MOVIE SCRIPT ACCEPTED BY A FILM STUDIO?

The Writers Guild of America Registration Office states that 20,000 scripts are registered with the Guild each year and that, of these, 1% are picked up by a studio and made into a film.

The Odds: **99.0-1** against

Source: Data from Writers Guild of America Registration Office.

WHAT ARE THE ODDS OF HAVING YOUR BOOK BECOME A BEST-SELLER?

First we decided what constitutes "best-sellerdom." There are all kinds of best-seller lists in the United States. The two that are generally accorded greatest respect by the publishing fraternity are the New York Times lists and those in the industry's bible, Publisher's Weekly. An analysis of these for 1978 shows that 126 different new titles hit the best-seller lists.

The Odds: **206-1** against

Source: New York Times and Publisher's Weekly.

WHAT ARE THE ODDS ON BACKING A HIT BROADWAY SHOW?

Broadway is booming and some shows are returning their backers' money plus handsome profits in less than 13 weeks. What's more, those profits can be big. For every $100 an investor put into "Fiddler on the Roof" the payback was $1,300. Nonetheless four out of five shows lose money.

The Odds: **4-1** against

Source: Data from Fortune Magazine.

WHAT ARE THE ODDS OF WINNING AN ACADEMY AWARD?

From the time these awards were started through 1978 there have been 7,147 nominations and 2,002 winners for odds of 2.6-1 against. However, the number of nominations is now normally five in each of 21 categories. Some nominations involve more than one person. In 1978 at the 51st ceremony 147 people were nominated.

The Odds: **6-1** against

Source: Academy of Motion Picture Arts and Sciences, Los Angeles, California.

WHAT ARE THE ODDS ON HAVING TO PAY THE I.R.S. MORE MONEY ON YOUR INDIVIDUAL TAX RETURN?

Of the 87,386,000 individual returns filed for fiscal 1978, a total of 7,567,313 drew added assessments (amounting to $448 million dollars). This included 91,000 returns paid by bum checks.

The Odds: against

Total	10.5-1
Failure to pay	22.1-1
Tax due	29.1-1
Delinquency	122.0-1
Bad check	959.0-1
Negligence	1,323.0-1
Other (includes not reporting tips)	12,119,110-1
Fraud	13,041,168-1

Source: Commission of Internal Revenue 1978 Annual Report, Department of the Treasury.

WHAT ARE THE ODDS THAT YOUR TAX RETURN WILL BE EXAMINED?

The I.R.S. examined 306,433 individual returns, 147,273 corporate returns and 27,579 partnership returns during fiscal 1978.

The Odds:	Individual	**284.0-1** against
	Corporate	**14.9-1** against
	Partnership	**42.7-1** against

Source: Ibid.

WHAT ARE THE ODDS OF BEING INDICTED OR CONVICTED FOR CRIMINAL TAX ACTIVITY?

Of the 136,718,000 tax returns of all kinds filed in fiscal 1978, just 2,634 cases were referred for prosecution and 1,724 indictments issued.

The Odds: against

Activity	
Referrals for prosecution	51,904-1
Indictments and informations	79,302-1
Plea of guilty or nolo contendere	114,985-1
Convicted after trial	607,634-1
Nol-prossed or dismissed	1,148,889-1
Acquitted	1,953,113-1

Source: Ibid.

WHAT ARE THE ODDS ON A CHECK BOUNCING?

According to the Federal Reserve Board, of the 32 billion checks written annually 0.67% bounce. It sounds like a small figure. It adds up to a large number of rubber checks, 214,000,000, almost one check for every man, woman and child in America.

Interestingly, Americans don't bounce checks where they most often (after banks) cash them: at the supermarket. The average supermarket cashes 1,707 checks each week yet just 1.5 bounce. What's more the average bounced check at $25.68 is well below the average of all checks cashed, $40.85. So if you're a supermarket operator the odds are 1,137-1 that any given check is O.K.

The Odds: **148.2-1** against

Source: Federal Reserve Bank and Food Marketing Institute, Loss Prevention Study.

WHAT ARE THE ODDS ON USING CREDIT CARDS BY FAMILY INCOME?

Just under two out of three Americans (64%) use credit cards. Usage is heaviest in the West (71%) and lowest in the East (61%). Understandably, the higher the family income the greater the use of credit cards.

The Odds: for unless noted	
Total	1.8-1
Income	
Under $10,000	1.5-1 against
$10,000-$14,999	1.9-1
$15,000-$19,999	1.9-1
$20,000 +	3.8-1

Source: The Gallup Report, Vol. 3, No. 3, March 1978.

WHAT ARE THE ODDS OF BEING PASSED A COUNTERFEIT BILL?

The odds are excellent if you work as a cashier in an establishment with a high turnover and rapid means of entering and exiting. For example, a fast food outlet is more likely to get hit than the local Mom and Pop shop or a sit-down restaurant. Most forgeries are so bad that even a cursory examination would spot the phonies.

In 1978 the Secret Service seized $18,300,000 worth of counterfeit money before it got into circulation while another $4 million was "successfully" passed. That word is in quotes advisedly. The capture and conviction rate for this particular crime is over 99%. The vast majority of counterfeits made are $10's and $20's and, less frequently, $50's. Seldom are larger denominations attempted. They attract attention, the one thing the passer doesn't want.

Counterfeiting hardly pays. Secret Service studies show the average profit among all counterfeiters is all of $40.

Due to its inferior quality a counterfeit bill tends to get spotted at a bank either by a teller or by an optical scanning device, so duds don't stay in circulation for long.

Assuming the average bill to be a $20 (the Treasury Department doesn't keep tabs on counterfeits by face value, which is surprising), 200,000 bills were in circulation in 1978. The U.S. Mint tells us that the total number of bills circulated during the year was 3,288,320,091. So, at worst, these are your chances of getting stung.

The Odds: **16,440-1** against

Source: U.S. Secret Service

WHAT ARE THE ODDS OF BEING ARRESTED IN A GAMBLING RAID FOR SELECTED CITIES?

If you have a hankering for the ponies and want to avoid the paddy wagon head for Tuscon. They didn't have a single arrest in 1973. Newark, on the other hand, had 1,356.

We thought the comparative results would be of interest. All these figures made the odds larger than they actually are since they're compared to the total population of each area.

The Odds: against

Newark	270-1
Cleveland	278-1
Kansas City, MO	302-1
St. Louis	308-1
Memphis	320-1
Honolulu	320-1
Dallas	354-1
Chicago	447-1
Fort Worth	449-1
Norfolk	489-1
Houston	495-1
Indianapolis	519-1
Los Angeles	548-1
Washington, D.C.	562-1
Philadelphia	576-1
Cincinatti	684-1
Louisville	702-1
Atlanta	719-1

The Odds: against

San Antonio	721-1
Tulsa	737-1
Baltimore	747-1
Rochester	841-1
Jacksonville	1,294-1
Jersey City	1,400-1
San Francisco	1,430-1
Boston	1,492-1
New Orleans	1,609-1
Tampa	1,671-1
Oklahoma City	1,677-1
Toledo	2,015-1
Austin	2,293-1
New York	2,325-1
Miami	2,363-1
Oakland	2,403-1
Omaha	2,403-1
Buffalo	2,832-1
Columbus, OH	3,256-1
Detroit	4,015-1
Seattle	4,081-1
Wichita	4,424-1
Akron	5,951-1
El Paso	6,409-1
Birmingham	7,041-1
Charlotte	8,620-1
Long Beach	9,900-1
Phoenix	11,363-1
San Diego	12,820-1
Portland, OR	14,820-1
Sacramento	15,624-1
San Jose	18,518-1
Denver	21,276-1
St. Paul	58,823-1
Minneapolis	66,666-1
Albuquerque	66,666-1
Milwaukee	111,110-1

Source: Commission on the Review of the National Policy Toward Gambling, *Gambling in America*, Washington, 1976.

BREADWINNERS ALL
At Work

Any form of work starts with being hired. And, at least in theory, getting a job depends on your ability to perform it well. That's why we started this chapter with a table of relative mental proficiency by state. There are two such uniform tests administered: the college entrance exams, which don't release state by state data, and the preinduction mental proficiency tests given to prospective draftees. This, along with the draft itself, has now gone by the boards. Even though it only tested males, it formed a measure of ability on a state by state basis.

If you live in the south you are justified in worrying about the quality of public education. By way of contrast our northernmost states seem to do exceptionally well right across America. Maybe those long cold winters provide a powerful incentive to hit the books.

Since we begin the chapter with a military test we continue on a martial note with the odds on reaching the top of the tree by branch of service.

Next we examine high school graduates' employment prospects by occupation. Predictably, remarkably few end up with any kind of authority, one reason why two out of three jump jobs within a year. That's one of the intriguing statistics in our next section which examines job mobility in detail. As you'll see women are far more stable employees than men. That's probably partly due to less opportunity to move, a phenomenon that's borne out by the small percentage of women (never as high as 30%) compared to men in the field of science.

Sex discrimination is one hindrance to getting ahead. We next look at the more frightening prospect of losing out because you've been injured.

We tell you how and where you're most apt to be working if you suffer an injury. Mining is still dangerous.

To end on an upbeat note we give you the odds on becoming president of a public company (they're daunting) and of being on the board (much better odds and the president reports to you).

WHAT ARE THE ODDS OF BEING COURTMARTIALED, BY BRANCH OF MILITARY SERVICE AND RANK?

In 1977 40 officers and 5,170 enlisted personnel underwent a general or special courtmartial. When compared to those in each branch of the service during that year several conclusions can be drawn. Naval officers live up to their gentlemanly reputation and Marines look for trouble.

In all services the probability of being convicted is exceptionally high. This ranges from a low of 78.3% for Army officers to a high of 100% for the very small number of Marine officers involved (three). With the exception of the Marines, officers get off more frequently than those in the ranks. Over all the odds of being convicted are 4.6-1 for.

The Odds: against

Army
Officers   4,216-1
Enlisted personnel   360-1

Air Force
Officers   10,666-1
Enlisted personnel   459-1

Navy
Officers   12,599-1
Enlisted personnel   491-1

Marine Corps
Officers   6,332-1
Enlisted personnel   133-1

Source: Department of Justice, *Sourcebook of Criminal Justice Statistics*, 1978.

IF YOU'RE IN THE MILITARY WHAT ARE THE ODDS, BY BRANCH OF SERVICE, ON REACHING THE TOP?

You're contemplating joining the military and you know you've got the makings of a general or admiral. Which branch should you join to hit the top? The numbers say the Air Force provides your best bet. Their total

uniformed manpower is 572,000 and they have 361 active duty gener-
als. As shown by the odds the Marines live up to their toughest outfit
reputation in promotions.

The Odds: against

Army
General	1,788-1
Brigadier General	3,568-1
Major General	4,561-1
Lieutenant General	21,416-1
Four Star General	77,099-1

Air Force
General	1,583-1
Brigadier General	3,176-1
Major General	4,399-1
Lieutenant General	15,051-1
Four Star General	43,999-1

Navy
Admiral—any	2,637-1
Rear Admiral	4,674-1
Vice Admiral	16,624-1
Admiral	59,110-1

Marines
General	2,908-1
Brigadier General	5,646-1
Major General	8,347-1
Lieutenant General	27,427-1
Four Star General	95,999-1

Source: Department of Defense Personnel, Pentagon.

**WHAT WERE THE ODDS OF A PROSPECTIVE DRAFTEE BEING EXAM-
INED, REJECTED AND INDUCTED DURING THE VIETNAM WAR?**

Almost 76 million men were classified for the draft. Of this total eight
million were examined, 3,880,000 rejected and 1,759,000 drafted into
service.

The Odds:	Examined	**7.8-1** against
	Rejected	**18.5-1** against
	Inducted	**42.0-1** against

Source: Department of Defense Personnel, Pentagon.

WHAT ARE THE ODDS ON A DRAFTEE FAILING TO MEET THE MENTAL REQUIREMENTS FOR INDUCTION INTO THE ARMED SERVICES?

While the draft is a thing of the past it revealed some interesting facts and figures on United States males. The 1972 mental ability failure rates were derived from a uniform test, providing a relative measure of the mental alertness and education level of men on a state by state basis.

In both Mississippi and South Carolina over one out of four draftees failed to meet minimum requirements. Georgia, Alabama and Louisiana also had very high rejection rates.

These five states are in sharp contrast to Minnesota, Montana, South Dakota, Wyoming, Washington, Oregon, New Hampshire, Nebraska and Kansas, all of which turned down two or fewer draftees per hundred.

The Odds: against

United States	13.9-1	Missouri	44.5-1
Alabama	4.5-1	Montana	82.3-1
Alaska	25.3-1	Nebraska	54.6-1
Arizona	21.7-1	Nevada	23.4-1
Arkansas	8.3-1	New Hampshire	49.0-1
California	21.7-1	New Jersey	26.0-1
Colorado	32.3-1	New Mexico	10.6-1
Connecticut	46.6-1	New York	17.9-1
Delaware	14.4-1	North Carolina	5.6-1
District of Columbia	7.0-1	North Dakota	46.6-1
Florida	11.5-1	Ohio	22.8-1
Georgia	4.1-1	Oklahoma	14.9-1
Hawaii	8.4-1	Oregon	49.0-1
Idaho	42.5-1	Pennsylvania	26.0-1
Illinois	19.8-1	Rhode Island	18.6-1
Indiana	20.1-1	South Carolina	2.9-1
Iowa	46.6-1	South Dakota	61.5-1
Kansas	51.8-1	Tennessee	6.4-1
Kentucky	9.0-1	Texas	11.0-1
Louisiana	4.8-1	Utah	33.5-1
Maine	33.5-1	Vermont	46.6-1
Maryland	19.8-1	Virginia	7.7-1
Massachusetts	40.7-1	Washington	54.6-1
Michigan	30.3-1	West Virginia	9.6-1
Minnesota	89.9-1	Wisconsin	30.3-1
Mississippi	2.8-1	Wyoming	54.6-1

Source: Department of the Army, Office of the Surgeon General, *Summary of Registrant Examinations for Induction.*

WHAT WERE THE ODDS OF BEING INJURED OR KILLED IN VIETNAM BY MILITARY BRANCH AND HOW DOES THIS COMPARE WITH PRIOR WARS?

If you or someone you cherished were in the Air Force the odds were unfortunately favorable. For the Marines, almost seven out of every 100 were killed or wounded, almost twice as many as in Korea. In every single service except the Air Force our Vietnam casualties per 100 combatants were higher than during the 1950-53 Korean War.

The Odds: against

Vietnam	Battle deaths	Other deaths	Total deaths	Wounds not mortal	Total casualties
Army	142-1	609-1	115-1	20.8-1	17.3-1
Navy	1,199-1	2,025-1	753-1	182-1	146-1
Marines	600-1	472-1	53.0-1	8.0-1	6.7-1
Air Force	1,298-1	2,885-1	895-1	502-1	321-1
Total	187-1	841-1	152-1	27.8-1	23.2-1
Korean War 1950-53					
Army	101-1	300-1	75.3-1	35.5-1	23.7-1
Navy	2,569-1	290-1	260-1	746-1	193-1
Marines	98.4-1	335-1	75.7-1	16.9-1	13.5-1
Air Force	1,070-1	217-1	180-1	3,491-1	171-1
Total	169-1	276-1	104-1	54.4-1	35.3-1
WWII 1941-46					
Army	46.9-1	134-1	34.4-1	18.9-1	11.7-1
Navy	112-1	162-1	65.8-1	110-1	40.7-1
Marines	32.9-1	139-1	26.3-1	9.0-1	6.3-1
Total	54.3-1	141-1	38.7-1	23.0-1	14.0-1

Source: Data from Department of Defense, Office of the Secretary.

WHAT ARE THE ODDS, BY SEX AND RACE, ON BEING EMPLOYED?

There is no sex discrimination when it comes to being out of work and looking for a job. The black unemployment rate is nearly twice that for whites.

The Odds: against

Women		Men	
White	11.7-1	White	14.6-1
Black	6.4-1	Black	6.9-1

Source: Department of Labor, Bureau of Labor Statistics, *U.S. Working Women: A Databook*, p. 44.

WHAT ARE THE ODDS ON MEN AND WOMEN WORKING AT LEAST PART OF THE YEAR, BY AGE?

Predictably enough, men start work earlier and work more years than women.

The Odds: for unless noted

Age	Men	Women
16 & 17	1.2-1	1.2-1 against
18 & 19	4.1-1	2.3-1
20-24	8.0-1	2.8-1
25-34	21.7-1	1.8-1
35-44	21.7-1	1.7-1
45-54	16.3-1	1.4-1
55-59	6.0-1	1.1-1
60-64	2.8-1	1.5-1 against
65-69	1.4-1 against	3.8-1 against
70 +	4.2-1 against	14.9-1 against

Source: Department of Labor, Bureau of Labor Statistics, *U.S. Working Women: A Databook*, p. 11.

WHAT ARE THE ODDS ON WORKING AT HOME?

Just 3.2% of employed Americans work at home.

The Odds: **30.2-1** against

Source: Department of Commerce, Bureau of the Census, *The Journey to Work in the United States*, No. 99, 1975, p. 23.

WHAT ARE THE ODDS ON COMMUTING BY TYPE OF TRANSPORTATION?

Energy crisis or no energy crisis, the vast majority of Americans (84.7%) continue to commute by car.

The Odds: against unless noted

Drive car alone	1.9-1 for
Car pool	4.6-1
Public transport	16.0-1
Walk	20.3-1
Bicycle	166-1
Motorcycle	249-1
Cab	499-1

Source: Department of Commerce, Bureau of the Census, *The Journey to Work in the United States*, No. 99, 1975, p. 23.

WHAT ARE THE ODDS ON COMMUTING TO WORK, BY ROUND TRIP DISTANCE?

Just one in 12 workers commute under two miles each day. Over one in three commute for 10 miles or more including one in 19 who commute over 50 miles a day.

The Odds: Less than 2 miles   7.1-1 against
           Less than 5 miles   2.5-1 against
           Less than 10 miles   2.0-1 for
           10 miles or more   2.0-1 against

Source: Ibid.

WHAT ARE THE ODDS ON MALE AND FEMALE HIGH SCHOOL GRADUATES NOT ENROLLED IN COLLEGE ENTERING DIFFERENT OCCUPATIONS?

Seven out of 10 such men start out as blue collar workers while very nearly six out of 10 women start out as white collar employees, with the vast majority of these doing some form of clerical work.

The Odds: against unless noted

	Males	Females
White Collar Workers	6.6-1	1.3-1 for
Professional, technical & kindred	14.2-1	89.9-1
Managers & administrators, except farm	54.6-1	89.9-1
Sales workers	29.2-1	8.3-1
Clerical workers	12.5-1	1.3-1
Blue Collar Workers	2.4-1 for	5.2-1
Craft & kindred	3.5-1	70.4-1
Operators except transport	3.4-1	5.8-1
Transport equipment operators	24.0-1	49.9-1
Laborers except farm and mine	3.6-1	34.7-1
Service Workers	8.7-1	3.2-1
Farm Workers	15.7-1	49.9-1

Source: U.S. Department of Labor, Bureau of Labor Statistics, *Special Labor Force Report No. 215, Students, Graduates & Dropouts,* 1977.

WHAT ARE THE ODDS ON MEN AND WOMEN GETTING THEIR JOBS IN VARIOUS WAYS?

For both sexes about one third of all workers get jobs they've applied for directly. Newspaper ads are next in importance for women, while men rely on friends to tell them about openings where the friends already work or elsewhere.

The Odds: against

	Women	Men
Applied directly to employer	1.9-1	1.8-1
Via friends	8.3-1	6.2-1
Via relatives	18.6-1	13.5-1
Answered newspaper ad	5.4-1	7.5-1
Employment agency (state or private)	6.6-1	10.4-1
Schol placement office	34.7-1	31.3-1
Union hiring hall	99.9-1	37.5-1
Civil service test	34.7-1	61.5-1

Source: U.S. Department of Labor, Bureau of Labor Statistics, *U.S. Working Women: A Databook,* p. 59.

WHAT ARE THE ODDS ON BEING EMPLOYED IN THE SAME JOB NEXT YEAR?

Farmers are the most stable group, while almost one out of five non-farm laborers will change jobs.

The Odds: for

	Male	Female
Total	7.7-1	7.5-1
Professional, technical and kindred workers	10.9-1	10.8-1
Managers and administrators, except farm	8.7-1	6.1-1
Sales workers	6.5-1	6.2-1
Clerical and kindred workers	6.6-1	6.8-1
Craft and kindred workers	8.4-1	6.0-1
Operatives, except transport	5.6-1	7.6-1
Transport equipment operatives	6.9-1	5.8-1
Laborers, except farm	3.9-1	5.1-1
Private household workers	——	10.0-1
Service workers, except private household	7.8-1	8.6-1
Farmers and farm managers	32.3-1	12.3-1
Farm laborers and supervisors	7.3-1	9.9-1

Source: Department of Labor News, (USD) 79-91.

WHAT ARE THE ODDS FOR MEN AND WOMEN, BY THEIR CURRENT AGE, OF HAVING THE SAME JOB A YEAR FROM NOW?

Two out of three 18 and 19-year-olds are not in the same job 12 months later. Job stability begins at 20; the older you get the more apt you are to stay put. Predictably women are less apt to stay with a job as they marry and have children. For example less than 3% of men between the ages of 25 and 34 are out of the work force, while the comparable number for women is 13.5%.

The Odds: for unless noted

Age	Men	Women
18-19	2.0-1 against	2.0-1 against
20-24	1.3-1	1.4-1
25-34	3.6-1	2.4-1
34-44	7.1-1	3.5-1
45-54	10.7-1	6.1-1
55-64	12.3-1	9.1-1
65 +	9.8-1	8.5-1

Source: Department of Labor, Bureau of Labor Statistics.

WHAT ARE THE ODDS OF HAVING THE SAME JOB A YEAR FROM NOW, BY DIFFERENT ETHNIC GROUPS OF MEN AND WOMEN?

Those who are most likely to stay with their present jobs are white males and black females.

The Odds: for

Men		Women	
White	4.5-1	White	2.8-1
Black	3.5-1	Black	3.6-1
Hispanic	3.5-1	Hispanic	2.2-1

Source: Department of Labor, Bureau of Labor Statistics.

WHAT ARE THE ODDS ON WHY MEN AND WOMEN WORK AT A SECOND JOB?

Understandably, about a third of both men and women take a second job to cover basic expenses. We're still a country with a strong work ethic. About one in five second job holders has another job for the sheer joy of it.

The Odds: against	**Women**	**Men**
Meet regular expenses	2.5-1	2.3-1
Enjoy the work	4.2-1	4.0-1
Save for the future	10.6-1	11.2-1
Held friend/relative	11.0-1	24.6-1
Buy something special	11.0-1	9.5-1
Get experience	14.9-1	13.9-1
Pay off debts	19.0-1	17.2-1
Other reasons	5.4-1	5.3-1

Source: Department of Labor, Bureau of Labor Statistics, *U.S. Working Women: A Databook*, p. 57.

WHAT ARE THE ODDS IN SELECTED FIELDS THAT AN EMPLOYEE WILL BE A WOMAN?

Women predominate in medical and health services (80%), local education (62%), service industries (63%), and finance, insurance and real estate (55%).

The Odds: against unless noted	
Total, nonagricultural industries	1.5-1
Private	1.6-1
Mining	13.3-1
Construction	13.3-1
Manufacturing	2.4-1
Durable goods	3.5-1
Non-durable goods	1.6-1
Transportation and public utilities	3.5-1
Communications	1.2-1
Wholesale and retail trade	1.4-1
Wholesale trade	3.2-1
Retail trade	1.1-1
Finance, insurance and real estate	1.2-1 for
Services	1.3-1 for
Personal	1.7-1 for
Miscellaneous business services	1.7-1
Medical and other health	4.0-1 for
Educational	Even
Government	1.2-1
Federal	2.4-1
State	1.3-1
State education	1.3-1

Other state government	1.2-1
Local	Even
Local education	1.6-1 for
Other local government	1.9-1

Source: Department of Labor, Bureau of Labor Statistics, *U.S. Working Women: A Databook*, p. 7.

WHAT ARE THE ODDS ON A WORKER BEING A WOMAN BY SELECTED OCCUPATION?

Today women account for 42% of professional technical workers and 20.8% of managers. Only 1.8% of engineers and 9.2% of doctors are female.

The Odds: against unless noted

Professional-Technical	1.4-1
Accountants	2.7-1
Engineers	54.6-1
Lawyers-judges	9.9-1
Physicians-osteopaths	6.8-1
Registered nurses	28.4-1 for
Teachers, except college & university	2.4-1 for
Teachers, college & university	2.2-1
Technicians, excluding medical-dental	6.4-1
Writers-artists-entertainers	1.9-1
Managerial-administrative, except farm	3.8-1
Bank officials-financial managers	3.0-1
Buyers-purchasing agents	3.2-1
Food service workers	1.9-1
Sales managers-department heads; retail trade	1.8-1
Clerical	3.7-1 for
Bank tellers	10.2-1 for
Bookkeepers	9.0-1 for
Cashiers	7.1-1 for
Office machine operators	2.8-1 for
Secretaries-typists	65.7-1 for
Shipping-receiving clerks	4.8-1

Source: Department of Labor, Bureau of Labor Statistics, *U.S. Working Women: A Databook*, p. 9.

WHAT ARE THE ODDS ON A WOMAN WORKING BY AGE?

Sixty-five percent of women aged 20-24 worked in 1976 and the figures are rising all the time. Well over half of all women up to the age of 55 are in the labor force.

The Odds: for unless noted

20-24	1.8-1
25-29	1.4-1
30-34	1.2-1
35-39	1.3-1
40-44	1.4-1
45-49	1.3-1
50-54	1.1-1
55-59	1.1-1 against
60-64	2.0-1 against
65-69	5.7-1 against
70 and over	20.7-1 against

Source: Department of Labor, Bureau of Labor Statistics, *U.S. Working Women: A Databook*, p. 60.

WHAT ARE THE ODDS OF A FEMALE SCIENTIST OR ENGINEER WORKING IN CERTAIN FIELDS COMPARED WITH A MAN?

We determined the ratio of females to males in each category; over one out of four psychologists is a woman, while only one in 250 is an engineer.

The Odds: against

Computer specialist	7.1-1
Engineer	249.0-1
Mathematical specialist	6.4-1
Life scientist	7.0-1
Physical scientist	12.3-1
Environmental scientist	30.3-1
Psychologist	2.6-1
Social scientist	4.3-1

Source: Department of Commerce, Bureau of the Census, *Current Population Report*, Series P-23, No. 76.

IF YOU'RE INJURED AT WORK WHAT ARE THE ODDS ON WHERE THE INJURY WILL OCCUR?

The National Safety Council estimates that in 1978 there were 2,200,000 disabling work accidents. Of this total 13,000 were fatal while 80,000 resulted in some permanent impairment.

The Odds: against

Eyes	15.6-1
Head (except eyes)	15.6-1
Arms	10.1-1
Trunk	2.7-1
Hands	15.6-1
Fingers	5.2-1
Legs	6.7-1
Feet	15.6-1
Toes	32.3-1
General	11.5-1

Source: Labor Department Reports.

WHAT ARE THE ODDS ON BEING KILLED AT WORK, OVER ALL AND BY INDUSTRY GROUP?

In 1978 there were a total of 13,000 deaths in a work force of 94,800,000. As shown by the odds, mining and quarrying is far and away the nation's most hazardous industry group.

The Odds: against

All industries	7,291-1
Trade	17,076-1
Service	13,470-1
Manufacturing	11,277-1
Government	9,058-1
Transportation & public utilities	3,399-1
Agriculture	1,841-1
Construction	1,768-1
Mining/quarrying	1,586-1

Source: National Safety Council, *Accident Facts,* 1979.

WHAT ARE THE ODDS ON A BUSINESS GOING BANKRUPT?

In 1978 a total of 6,619 businesses failed. The average debt when they went under was a hefty $355,946. The failure rate per 10,000 businesses has varied dramatically over the last five years as shown by the figures which follow.

The Odds: against

1978	416-1
1977	356-1
1976	285-1
1975	232-1
1974	262-1

Source: Dun and Bradstreet, Inc., New York City.

WHAT ARE THE ODDS ON BECOMING PRESIDENT OF A PUBLICLY OWNED COMPANY?

The Securities and Exchange Commission requires 9,263 firms to file 10K reports as publicly held. In total we estimate that these firms employ 21,300,000.

The Odds: **2,298-1** against

Source: S.E.C. and Heron House estimate.

WHAT ARE THE ODDS OF BEING ON THE BOARD OF A PUBLICLY OWNED COMPANY?

We took the figure from Standard and Poor's register of 8,000 companies and adjusted it upward, assuming that other companies have the same average number of directors (8.9). We presumed that all directors of publicly held companies are 18 or older.

The Odds: **1,839-1** against

Source: Standard and Poor's and Heron House estimate.

CASTLES, CONTENTS AND COMMUNICATIONS
At Home and Keeping in Touch

"Give me land, lots of land, under starry skies above, don't fence me in" is no longer the average American's credo. Quite simply that land is now too expensive (see millionaires under *For the Love of Money*). Today for every 100 houses started just 16 are built on plots of a quarter acre or more. As one leading official of a realty group puts it, "It's no longer home on the range. Today the name of our game is home on the hot plate."

This hasn't deterred people from seeking mortgages (one out of two succeed) or from becoming home-owners. But, as you'll see in the chapter, many more whites than blacks own their homes. Nothing has risen faster than the cost of housing and that's reflected in our odds on a home being worth certain amounts. Houses in the $35,000 to $50,000 category form the largest group.

If you've ever answered an ad for a 3½ room apartment only to discover that one room was a hall the size of a closet and another a four foot alcove, our odds on home size should be of interest. Home-owners have very nearly twice as much room as renters.

Next we turn our attention to electrical appliances. It won't come as electrifying news to learn that we virtually all own TV sets, toasters, radios and refrigerators. But it is of interest to see that advertising and editorial comments aside, microwaves and food processors have a way to go before they reach universal acceptance.

As long as we're discussing items around the house we'll next look at those which put your family at risk. There are some surprises here. A first bicycle under the Christmas tree is always a momentous occasion for both child and parents. The accidents which follow are equally momentous. Almost half a million bicycle related accident victims show up in hospital emergency rooms each year.

Next we look at a variety of other things that may or may not feature in your surroundings. More than likely man's best friend is your constant companion; two out of three households have a dog. You're far less likely to be in the swim of things, as shown by your swimming pool odds.

Finally we address ourselves to your mail and your telephone. You'll discover that the check may well be lost in the mail but if so it's probably been misaddressed. You'll also see the Post Office statistics on speed of delivery. We don't believe these figures either.

WHAT ARE THE ODDS ON SELLING YOUR FRIENDS SHORT?

The total sales force in retail clothing establishments is 905,400 and it's estimated that half this number sell men's clothing (including shorts). Across America about 1.2% of sales are made to people who are "personally well known to the salesperson." From the above we assembled the odds on selling your friends shorts.

The Odds: **47,522-1** against

Source: Bureau of Labor Statistics and confidential retail studies.

WHAT ARE THE ODDS OF FEELING YOUR OATS?

At the time of the latest Agriculture Census 358,124 farms reported that they had grain under cultivation. These folks can fairly be said to be the only ones who feel their own oats when compared to all households.

The Odds: **194-1** against

Source: Department of Commerce, Bureau of the Census, *Statistical Abstract of the United States*, 1979.

WHAT ARE THE ODDS ON A LEFT-HANDED COMPLIMENT?

It is difficult to determine what percentage of the public is left-handed, and whom therefore tender compliments while possessing this trait. Our best guess—12%—is based on the sale of baseball gloves exclusive of first base where southpaws are preferred.

The Odds: **7.3-1** against

Source: Data from Wilson Sporting Goods.

WHAT ARE THE ODDS OF GETTING A MORTGAGE?

One out of every two mortgage applicants succeeds.

The Odds: **Even**

Source: Department of Agriculture, Farmers Home Administration.

WHAT ARE THE ODDS OF OWNING YOUR HOME?

In 1977, the latest year for which figures were available of the United States' 75,280,000 dwelling units, 38,754,000 were owner-occupied. Thus the odds are slightly better than even.

The Odds: **1.1-1** for

Source: Department of Commerce, Bureau of the Census, *Statistical Abstract of the United States,* 1979.

WHAT ARE THE ODDS OF OWNING YOUR HOME BY RACE AND REGION OF THE COUNTRY?

In 1976 a total of 67.6% of white citizens lived in their own homes compared to 38.4% of blacks.

The Odds:

	Whites	Blacks
Total United States	2.1-1 for	1.3-1 against
Northeast	1.7-1 for	2.3-1 against
North Central	2.5-1 for	Even
South	2.3-1 for	1.1-1 against
West	1.7-1 for	1.4-1 against

Source: Department of Commerce, Bureau of the Census, *Statistical Abstract of the United States,* 1978.

WHAT ARE THE ODDS ON YOUR HOME BEING WORTH PARTICULAR AMOUNTS?

In 1977 less than 17% of all owner-occupied homes were worth under $20,000 while 28% of homes were worth over $50,000. The rise in home values has been startling over the last seven years. With an average of less than 14 million new home starts in all price categories over the period, the number of $50,000+ owner-occupied homes soared from 997,000 to 10,862,000.

The Odds: against unless noted

Less than $20,000	5.0-1
Less than $24,999	3.0-1
Less than $34,999	1.2-1
Less than $49,999	2.6-1 for
$50,000 or more	2.6-1

Source: Department of Commerce, Bureau of the Census, *Statistical Abstract of the United States*, 1979.

WHAT ARE THE ODDS ON HAVING A HOME ON AT LEAST A QUARTER OF AN ACRE?

If you're in an older home they're much better; 22.4% of families in these have at least a quarter of an acre. Yet of all new housing starts in 1977, just 16% have this much space.

The Odds:	New homes in 1977	**5.2-1** against
	Older homes	**3.5-1** against

Source: Department of Housing and Urban Development—FHA, *Series Data Handbook*, Section 203B, "Home Mortgage Statistics," 1978, p. 62.

WHAT ARE THE ODDS ON YOUR HOME HAVING A SPECIFIED NUMBER OF ROOMS BY OWNER/OCCUPIERS VERSUS RENTERS?

Home-owners clearly have far more space than renters: a median of 5.8 rooms versus 4.1 for those paying rent. Fully 68% of home-owners/occupiers have six or more rooms while only 16% of renters can claim this amount of space. By way of contrast only 2% of homeowners have less than four rooms; the equivalent figure for renters is 31%, almost one out of three.

The Odds: against

	Owner/occupiers	Renters
1 room	——	33.2-1
2 rooms	49.0-1	13.3-1
3 rooms	——	3.8-1
4 rooms	7.3-1	1.9-1
5 rooms	2.6-1	4.3-1
6 rooms	2.7-1	9.0-1
7 or more	2.2-1	15.7-1

Source: Department of Commerce, Bureau of the Census, *Statistical Abstract of the United States*, 1978.

WHAT ARE THE ODDS ON HAVING TWO OR MORE BATHROOMS?

According to the Department of Housing and Urban Development, 44.1% of existing homes have one and a half to two bathrooms (those halves being just a john or just a shower) while 2% of homes have more than two full bathrooms.

The Odds: **1.2-1** against

Source: Department of Housing and Urban Development, FHA, *Series Data Handbook*, Section 203B, "Home Mortgage Characteristics," 1978, p. 10.

WHAT ARE THE ODDS ON MOVING IN ANY GIVEN YEAR?

About one third of our population has made a move in any given two year period. Of those who use a moving company one third move to another state.

The Odds: **4.7-1** against

Source: *The Gallup Report*, Vol. 2, No. 4, 1978.

WHAT ARE THE ODDS ON OWNING CERTAIN ELECTRICAL APPLIANCES?

All homes have certain basics. Radios, refrigerators, TV sets, toasters, irons and vacuum cleaners are in 99.9% of homes wired for electricity (though not necessarily in working order as many would grumpily point out). So the odds on owning these are overwhelming. Ownership of other appliances can vary widely. Microwave ovens are a good example. Only 6.7% of households had one in 1977, the latest year for which figures are available. Here are the odds which you may find useful in selling ("John, over eight out of 10 families have color TV") or stonewalling ("Over half the population do without food disposers; we can too").

The Odds: for unless noted

Room air-conditioner	1.3-1
Blender	Even
Can-opener	1.4-1
Clothes washer	2.8-1
Clothes dryer	1.5-1
Dishwasher	1.4-1 against
Food disposer	1.3-1 against
Freezer	1.2-1
Microwave oven	13.9-1 against
Mixer	12.1-1

Source: *Merchandising Annual Statistics Number*, Gralla.

WHAT ARE THE ODDS OF BEING INJURED BY VARIOUS ITEMS AROUND THE HOUSE?

The next time someone with a shiner gives you the classic "I ran into a door" bit he may be telling the truth. One out of 360 families suffered a door-related injury in 1977. As the figures show our homes are veritable booby traps. Even your cocktail table is out to get you. These accounted for almost 64,000 injuries, more than all ladders. Stairs and steps are the biggest danger around the house, accounting for over half a million visits to the emergency room each year. As the figures show, bikes are also a prescription for trouble, causing nearly half a million injuries.

All these figures are compiled by the Consumer Products Safety Commission which takes a representative sample of emergency room accidents and projects it up to a national total.

Some of the totals are hardly surprising. Most of us know that skateboards, saws and lawnmowers can be dangerous. But other figures give one pause for reflection. It seems unfair to be done in while practicing safety, yet both lightning rods and fire extinguishers take their toll. Who's to say "I'd be better off if I'd stayed in bed" when beds caused 77,581 injuries and pillows another 445. Exercise isn't necessarily healthy and sex can lead to a trip to the hospital emergency room. According to these facts the odds are 47,468-1 that even this book may do you in!

The Odds: against

Stairs and steps	118-1	Cocktail tables	1,156-1
Bicycle	149-1	Ladders	1,194-1
Football equipment	181-1	Playground climbing	1,645-1
Baseball	184-1	Lawnmower	1,722-1
Basketball	198-1	Slides	2,965-1
Glass fixture	205-1	Trampolines	4,115-1
Nails, carpet tacks &		Exercise equipment	4,630-1
screws	249-1	Nail preparation	18,524-1
Knives and pocketknives	332-1	Books, magazines &	
Doors	360-1	albums	47,468-1
Skate boards	528-1	Contraceptive devices	50,891-1
Unupholstered chairs	649-1	Lightning arresters	71,524-1
Tables	779-1	Fire extinguishers	85,762-1
Glass	832-1	Pillows	166,515-1
Saws (manual & powered)	881-1	Bubble baths	1,323,213-1
Swings	1,024-1	Burglar alarms	3,899,999-1
Tub and shower	1,028-1	Pacifier	12,349,999-1

Source: Consumer Products Safety Commission, *Tabulation of Data from National Electronic Injury Surveillance System*, 1977.

WHAT ARE THE ODDS ON HAVING A HANDGUN IN YOUR HOME?

Various local and statewide audits tend to indicate that 20% of all households have at least one handgun on the premises.

The Odds: **4.0-1** against

Source: Data from Department of Justice.

WHAT ARE THE ODDS OF SUBSCRIBING TO CABLE TV?

In 1977 11,900,000 households subscribed to cable TV.

The Odds: **5.2-1** against

Source: Department of Commerce, *Statistical Abstract of the United States,* 1978.

WHAT ARE THE ODDS OF HAVING AN AX TO GRIND?

According to the True Temper Company there's at least one ax in every second house.

The Odds: **Even**

WHAT ARE THE ODDS OF LIVING IN A SKYSCRAPER?

Oddly we haven't been able to get this precisely. The number of buildings put up in the last five years (through 1978) of over four stories is as close as you can come. It's just 2,700 buildings. Over the same period of time 3,000 buildings were constructed with over 50 units (but this includes sprawling condo types). The best guestimate as to the number of residential skyscrapers of over 15 stories is 800 nationwide.

The Odds: **662-1** against

Source: Department of Commerce, Bureau of the Census, Construction Statistics Division.

WHAT ARE THE ODDS ON OWNING VARIOUS PETS?

Fifty million American families own a dog and one out of three of these has a second dog or a cat or both. Half this number, 25 million, own cats, 8% of all households own fish and 5% own birds.

The Odds:	Dog	**2.0-1**	for
	Cat	**2.0-1**	against
	Fish	**11.5-1**	against
	Birds	**19.0-1**	against

Source: *Pet Supplies Magazine,* Duluth, Minnesota.

WHAT ARE THE ODDS OF HAVING YOUR OWN SWIMMING POOL?

As of January 1, 1979 there were 1,424,900 residential in-the-ground swimming pools in the United States, according to the National Swimming Pool Institute. Exact statistics on above ground pools are impossible to come by but the same source estimates there are 2,500,000 of these.

	The Odds:	A swimming pool	**18.2-1**	against	
		An in-ground pool	**52.0-1**	against	
		An above ground pool	**29.2-1**	against	

Source: Data from the National Swimming Pool Institute.

WHAT ARE THE ODDS OF OWNING A LAWN TRACTOR?

According to the Outdoor Power Equipment Institute, 750,000 units were shipped in 1977. John Deere estimates that 5% of home-owners have lawn tractors, and report that ownership is highest in New England and the Midwest (Indiana is number one in per capita ownership).

The Odds: **29.8-1** against

Sources: Outdoor Power Equipment Institute, Washington, D.C. and John Deere and Company, Moline, Illinois.

WHAT ARE THE ODDS OF HAVING A PRIVATE TENNIS COURT?

According to the U.S. Tennis Court and Track Builders Association there are 150,000 courts in America today. The U.S. Tennis Association estimates that 18,000 of them are at private homes.

The Odds: **4,193-1** against

Sources: Data from U.S. Tennis Court and Track Builders Association, Glenview, Illinois and U.S. Tennis Association Tennis Statistics, 1977, New York.

WHAT ARE THE ODDS OF HAVING A COVERED GARAGE?

Of all homes in 1977 just over one in four had no garage, one in about five has a carport and the rest had true garages.

The Odds: **1.3-1** for

Source: Department of Commerce, *Statistical Abstract of the United States*, 1978, p. 788.

WHAT ARE THE ODDS ON MISSING A LETTER?

Of the 56 billion pieces of mail handled by the U.S. Postal Service in 1978, 4.5 billion were not delivered as addressed.

The Odds: **11.4-1** against

Source: U.S. Postal Service

WHAT ARE THE ODDS ON A LETTER BEING LOST IN THE MAIL?

In 1978 211,203 letters were reported lost. Clearly more were lost but the Postal Service has no further data on this topic.

The Odds: **264,140-1** against

Source: U.S. Postal Service

WHAT ARE THE ODDS ON A LETTER ARRIVING WITHIN A CERTAIN PERIOD OF TIME?

Assuming the letter is properly addressed and posted before 5 P.M., these are the odds by distance and time on a letter arriving.

The Odds: Less than 50 miles/within   **19.0-1** for
24 hours
50 to 600 miles/within 48 hours   **6.1-1** for
600 + miles but within conti-   **8.1-1** for
nental U.S./within 72 hours

Source: U.S. Postal Service

WHAT ARE THE ODDS ON RECEIVING JUNK MAIL ON ANY GIVEN DAY?

About 19% of American households receive at least one piece of junk mail a day. Just over half of these (51%) would prefer not to receive such literature.

The Odds: **4.3-1** against

Source: *The Gallup Report*, Vol. 2, No. 3, 1978.

WHAT ARE THE ODDS ON HAVING A TELEPHONE?

In 1976 there were 114,491,000 residential phones. Reasonably enough Washington, D.C. is at the top of the telephone user poll; there are

145.8 telephones there for every 100 people. You're least apt to reach someone in Hammond, Indiana where there are just 50.8 phones for every 100 residents. Over all, 95% of homes have a phone.

The Odds: **19.0-1** for

Source: Federal Communications Commission, Washington, D.C.

WHAT ARE THE ODDS THAT YOU'LL BE ON THE TELEPHONE IN A GIVEN HOUR ON A WEEKDAY?

Incredibly good. Assuming that phone calls are rarely placed or received by children under five years old, there were about 200 million Americans in 1979 placing and receiving 190 billion 222 million phone calls. That's 901 phone calls for every man, woman and child (over five) per year, or about two and a half phone calls per day. Sixty-three percent of these are made during weekday business hours.

The Odds: **7.3-1** against

Source: *AT&T Annual Report*, 1979.

WHAT ARE THE ODDS THAT YOU'LL GET A BUSY SIGNAL?

Obviously it depends on who you call. The odds are based on a Bell Labs study of long distance direct dial calls.

The Odds: **8.9-1** against

Source: Bell System, *Technical Journal*, Vol. 57, No. 1, 1978.

WHAT ARE THE ODDS ON NO ONE ANSWERING THE PHONE?

Over all, 12.7% of calls go unanswered.

The Odds: **6.9-1** against

Source: Bell Labs System, *Technical Journal*, Vol. 57, No. 1, 1978.

WHAT ARE THE ODDS ON MAKING AT LEAST ONE LONG DISTANCE CALL IN THE COURSE OF A YEAR?

Over nine out of 10 customers do.

The Odds: **14.4-1** for

Source: *AT&T Annual Report*.

Chapter

JOCKS, JOGGING AND JUMPS
At Play

A look out your window on any given morning would lead you to believe that only the family that jogs together stays together. No so. Only 11% of Americans hit the streets. But then as a nation we're hardly exercise freaks; less than half of us exercise at all. Incidentally, many maintain that Democrats tend to get exercises more than Republicans. It's true.

Every sportsman strives for perfection, so we're giving you the odds of making a hole inone and bowling a perfect game. In each case these are long but not nearly as long if you're a pro. On holes in one pros do 10 times as well as amateurs, while they're 100 times more successful at bowling 300.

Professionals may do better in these respects but they still encounter long odds on making bit money on tour. For both golf and tennis only one in about 15 pros makes $100,000 or more.

The odds of success are markedly shorter for our most gifted amateurs, our Summer Olympic contestants. They pick up one out of every four medals. In Winter Olympic sports such as skiing it's all downhill. We don't do nearly as well.

For aspiring athletes we profile the odds of starring on various high school and college teams, and for worried parents we outline the odds on being injured by sport. Football is dangerous. The odds on suffering permanent serious injury (1,581-4) may sound long. That's still one man for every 18 complete football teams.

Sideline spectacles have long been a colorful part of our sports scene so we present the odds on making one of sport's greatest shows, the Dallas Cowboy cheerleaders. Being chosen is like the league itself, getting tougher all the time.

Finally, for that minority of readers who simply love to fling themselves from planes we offer the facts on malfunctioning parachutes. As you'll see square chutes are infinitely to be preferred. Happy landings.

WHAT ARE THE ODDS OF BEING BEHIND THE EIGHT BALL?

The leading journal in the billiards field reports that an estimated 10 million different people play pool in any given year.

The Odds: **20.9-1** against

Source: Billiards Digest

WHAT ARE THE ODDS THAT IT'S A SURE BET?

According to the Jockey Club several horses have been named Sure or Sure Bet. The first, foaled in 1918, was called Sure. Sadly, no records on the horse exist. The second, a filly also named Sure, was foaled in 1951 but never raced. This was also the case with a filly named Sure Bet foaled in 1953. Another filly foaled in 1969 and named Sure Bet ran eight times without winning. Finally, another filly named Sure Bet was foaled in 1977 and has yet to race as we go to press. So the odds on a Sure Bet being a filly are excellent. These are the odds of winning with a Sure Bet based on the races so far.

The Odds: **8.0-1** against

Source: American Jockey Club

WHAT ARE THE ODDS THAT YOU'LL GO FLY A KITE?

In 1978 some 150 million kites were sold in the United States. Clearly some purchasers bought more than one kite, so the odds we came up with (based on all those over the age of five at the time) are somewhat larger than shown. Nonetheless when you tell someone to go fly a kite the probability of their doing so is strong.

The Odds: **3.8-1** for

Source: *Time*, June 12, 1978.

WHAT ARE THE ODDS OF RAISING A RACKET?

In 1976, 29,210,000 different people played at least one game of tennis. Comparison with the total population over the age of five gives us the odds on those who raised a racquet at least once.

The Odds: **5.7-1** against

Source: U.S. Tennis Association

WHAT ARE THE ODDS ON PARTICIPATING IN VARIOUS SPORTS?

The continued popularity of jogging shows up in our odds: two out of three Americans over the age of 12 participate in the activity.

The Odds: against unless noted

Walking/jogging	2.1-1 for	
Swimming	1.7-1 for	
Boating	1.4-1 for	
Fishing	1.1-1 for	
Camping	Even	
Bicycling	1.2-1	
Tennis (outdoor)	2.1-1	
Hiking/backpacking	2.6-1	
Hunting	4.3-1	
Golf	5.7-1	
Horseback riding	6.1-1	

Source: Data from U.S. Heritage, Conservation and Recreation Service.

WHAT ARE THE ODDS THAT YOU EXERCISE DAILY?

According to a 1978 Gallup poll 47% of the population exercise daily. That's a startling increase on 1961's 24%. Who's least apt to exercise? Those with only a grade school education (30%), those earning between $3,000 and $5,000, older people and Republicans.

The Odds: **1.1-1** against

Source: Gallup Poll, February 1978.

WHAT ARE THE ODDS THAT YOU JOG AND IF SO HOW FAR?

Just under 11% of the population was jogging in 1978, a figure that doubtless has gone up. Of these, 23% jog for under one mile, 37% over one mile but less than two, and 23% over two miles but less than three. A hardy 14% jog over three miles per day.

The Odds: On Jogging **8.2-1** against

On Distance Jogged Among Joggers
Under 2 miles **1.5-1** for
Over 2 miles   **1.5-1** against

Source: Gallup Poll, February 1978.

WHAT ARE THE ODDS OF SCORING A HOLE IN ONE?

Golf Digest's hole in one clearing house recorded 28,576 holes in one during 1978. Maude Muth, at 86, became the oldest woman to score a hole in one when she sank her five wood tee shot on the 102 yard 14th hole at her home course, Kings Inn Golf and Country Club in Florida. It was her first hole in one in 50 years of golf.

Here are the odds according to the magazine's computations for golfers playing over a standard 18 hole course.

The Odds:	Regular golfer	**10,738** against	
	PGA Tournament player	**927-1** against	

Source: Data from *Golf Digest.*

WHAT ARE THE ODDS OF BOWLING A PERFECT GAME?

We talked to the American Bowling Congress, the Professional Bowlers Association, the Women's International Bowling Congress and the Women's Pro Bowling Association. The odds below are interesting for several reasons. They reveal that bowlers clutch in tournament play, that pros are light years better than amateurs and that women are far poorer bowlers than men.

The Odds: against

Men in league play	420,000-1
Men in tournament play	846,000-1
Men (pro) in tournament play	8,000-1
Women in league play	18,000,000-1
Women (pro) in tournament play	125,000-1

Source: Data from American Bowling Congress, Professional Bowlers Association, The Women's International Bowling Congress and the Women's Pro Bowling Association.

WHAT ARE THE ODDS ON BEING ABLE TO SWIM?

According to the Red Cross one third of the population over the age of 14 can't swim.

The Odds: **2.0-1** for

Source: American Red Cross

WHAT ARE THE ODDS OF BEATING YOUR OPPONENT IN STRAIGHT SETS?

The Association of Tennis Professionals gives the following figures based on their members' play in the United States Open Tennis Championship over the past five years.

The Odds:	Men's singles	**2.3-1** against	
	Men's doubles	**2.3-1** against	
	Women's singles	**2.3-1** against	
	Women's doubles	**2.3-1** against	

Source: Data from Association of Tennis Professionals.

WHAT ARE THE ODDS OF WINNING OVER $100,000 IN A GIVEN YEAR AS A MALE TENNIS PROFESSIONAL?

In 1978 a total of 492 players earned money in male professional tennis. Of these, according to the Association of Tennis Professionals, 34 won $100,000 or more.

The Odds: **14-1** against

Source: Ibid.

WHAT ARE THE ODDS OF WINNING A P.G.A. GOLF TOURNAMENT?

The average number of golfers who enter a tournament is 150. There is, of course, just one winner.

The Odds: **150-1** against

Source: Data from Professional Golfers Association.

WHAT ARE THE ODDS OF WINNING $100,000 ON THE P.G.A. TOUR?

The P.G.A. reports there were 370 P.G.A. members who were eligible to compete in the tour in 1978. Of these 24 won over $100,000.

The Odds: **15-1** against

Source: Ibid.

WHAT ARE THE ODDS OF WINNING AN OLYMPIC MEDAL?

In the 1976 Olympics 529 Americans were entered. We won 112 medals, 21% of all medals awarded. This varied widely between the

Summer Olympics where we did well: 402 Americans entered, 96 medals won (24% of the total); and the Winter Olympics where we did poorly: 127 Americans entered, six medals won (4% of the total). These are the odds on an American athlete winning an Olympic medal.

The Odds: Overall   **3.7-1** against
            Summer medal   **3.2-1** against
            Winter medal   **20.1-1** against

Source: Data from the Olympic Committee

WHAT ARE THE ODDS ON BEING IN YOUR HIGH SCHOOL'S STARTING LINEUP IN BASEBALL, BASKETBALL AND FOOTBALL?

According to the High School Athletic Coaches Association 13,500 schools have boys' baseball teams and 425,000 boys play, 19,600 schools have boys' basketball teams with 760,000 participants, while 14,000 schools have girls' basketball teams with 650,000 girls playing. Finally, there are 1,100,000 football players on 15,000 teams.

		Boys	**Girls**
The Odds:	Football	**3.0-1** against	——
	Basketball	**8.0-1** against	**9.0-1** against
	Baseball	**3.5-1** against	——

Source: Data from High School Athletic Coaches Association

WHAT ARE THE ODDS OF BEING SELECTED TO PLAY PRO BASE-BALL WHILE STILL IN HIGH SCHOOL?

It's reported that 300 of the 500 players last drafted were coming directly from high school (the rest were from colleges). The odds are based on the 425,000 high school baseball players.

The Odds: **1,416-1** against

Source: Data from High School Athletic Coaches Association

WHAT ARE THE ODDS OF A MAN MAKING THE STARTING LINEUP FOR COLLEGE FOOTBALL, BASKETBALL AND BASEBALL?

Four hundred and eighty colleges play football with average rosters of 90 players, and 22 in the offensive and defensive starting lineups. Seven hundred schools play basketball with an average squad size of 15, and 607 colleges play baseball with an average team size of 20.

The Odds:	Football	**3.0-1** against
	Basketball	**2.0-1** against
	Baseball	**1.2-1** against

Source: National College Athletic Association

WHAT ARE THE ODDS ON BEING KILLED OR SERIOUSLY INJURED WHILE PLAYING FOOTBALL?

The answer according to the Journal of the American Medical Association is "far better than it used to be." From 1959-1963 there were 86 deaths among 820,000 players. So the odds for that period were one in 9,534 players.

With the arrival of the face headmask these figures have dropped. Still in 1975 there were an average of 15 deaths and 161 serious permanent injuries each year. Of all paralyzing accidents over 70% occur to the person that's doing the tackling. The figures chart 80% for college players. The most risky position to play both in terms of death and injury: defensive back.

The Odds:	Being killed	**16,558-1** against
	Suffering permanent serious injury	**1,584-1** against
	That the serious injury will involve a neck fracture	**Even**

Source: Data from *Journal of the American Medical Association.*

WHAT ARE THE ODDS OF A MAN BEING HURT IN VARIOUS COLLEGE SPORTS?

When it comes to college sports, wrestling has a hammerlock on the injuries. Three out of 10 participants are out of action for at least a week or require dental work as a result of the sport.

The Odds: against

Wrestling	2.4-1
Football (fall)	3.2-1
Ice hockey	3.7-1
Basketball	5.0-1
Volleyball	5.0-1

The Odds: against

Tennis	7.0-1
Soccer	7.2-1
Football (spring)	8.5-1
Cross country	9.3-1
Gymnastics	9.6-1
Lacrosse	10.0-1
Indoor track & field	10.2-1
Outdoor track & field	11.0-1
Baseball	13.7-1
Swimming and diving	44.4-1

Source: National Athletic Injury/Illness Reporting System, Pennsylvania State University.

WHAT ARE THE ODDS OF A WOMAN BEING HURT IN VARIOUS COLLEGE SPORTS?

Gymnastics generates just as many injuries for women as wrestling does for men: one out of three participants gets injured. Badminton isn't normally regarded as a hazardous pastime, yet 16.6% of players are laid up as a result of the sport.

The Odds: against

Gymnastics	2.4-1
Badminton	5.0-1
Basketball	6.4-1
Cross country	6.4-1
Squash	7.0-1
Indoor track and field	9.6-1
Volleyball	9.7-1
Tennis	17.2-1
Softball	18.2-1
Outdoor track and field	19.8-1
Field hockey	23.4-1
Lacrosse	23.4-1
Swimming and diving	39.0-1

Source: National Safety Council

WHAT ARE THE ODDS OF BECOMING A DALLAS COWBOYS CHEERLEADER?

According to the Dallas Cowboys they're becoming longer all the time. There is no residency requirement for these tryouts, so anyone can enter.

You don't even have to know the rudiments of the game. Open tryouts have been held since 1976 when 250 women applied and 32 were accepted. Since then the number of cheerleaders has gone up marginally to 36 but the number of applicants has increased more than sixfold.

The Odds: 1976   **8.0-1** against
           1977   **19.0-1** against
           1978   **29.0-1** against
           1979   **45.0-1** against

On the basis of the past we predict that 2,280 applicants will try out in 1980 and only 36 will make the cut.

The Odds: 1980   **63.0-1** against

Source: Data from the Dallas Cowboys Cheerleaders.

WHAT ARE THE ODDS ON A DRAFT PICK MAKING AN N.F.L. TEAM?

First round picks are assured of playing. This holds true right through the fourth round choices. Thereafter the tide begins to turn so that even if the Pittsburgh Steelers pick you up in the 15th round, you'd be better off becoming an insurance salesman.

The Odds: for unless noted

Draft round	
1st	100-1
2nd	10-1
3rd-4th	5-1
5th-7th	5-4
8th-10th	4-3 against
11th-14th	7-3 against
15th or lower	3-1 against

Source: Sports Products, Inc., Norwalk, Connecticut.

WHAT ARE THE ODDS ON WINNING BY VARIOUS MARGINS AND OVERTIME IN PRO FOOTBALL?

If you play point spreads here's a handy quide to total points scored and winning margins. It shows that most pro games are hard fought, close contests.

The Odds: against

The winning team

Scoring 30 + points	2.0-1
Scoring 40 + points	11.0-1
Scoring 50 + points	100.0-1
Winning by 6 points or less	2.2-1
Winning by 14 points or less	3.0-2
Going into overtime	45.0-1

Source: Ibid.

WHAT ARE THE ODDS OF WINNING AGAINST ALL ODDS IN PRO FOOTBALL?

It's late in the fourth quarter and your team's down. Your opponent offers to double his bet. Should you accept? Even if your team seems to be sitting on its feet, the odds below may tell you how far to gamble.

The Odds: against

Winning

While losing in the third quarter	5.5-1
While losing with two minutes to play	8.5-1
After scoring fewer touchdowns	32.0-1
After gaining fewer yards	3.0-1
After losing more turnovers	4.5-1
After making fewer first downs	12.0-1
After running fewer plays from scrimmage	9.0-5

Source: Ibid.

WHAT ARE THE ODDS OF A SUDDEN CHANGE IN A PRO FOOTBALL GAME?

Your N.F.L. team's doing well when disaster strikes. The odds show that the least anticipated dangers, fumbles and interceptions, are most likely to swiftly turn the tables. (All the odds that follow are for a single team.)

The Odds: against

Fumble recovered/returned for a touchdown	7-1
Interception returned for a touchdown	7-1
At least one extra point missed	11-4
Punt returned for touchdown	20-1
Kickoff returned for touchdown	45-1

Source: Ibid.

WHAT ARE THE ODDS ON VARIOUS TYPES OF PERFECT GRIDIRON PERFORMANCE?

Every N.F.L. coach strives for perfection. How attainable is it? Hard to achieve is the answer. Shutting out an opponent only occurs in one out of six games. Fumbles do occur and interceptions are the rule rather than the exception. As the odds indicate, most coaches will go a lifetime without an undefeated season. (All the odds that follow are for a single team.)

The Odds: against

Allowing the opponent no touch downs	5.0-1
No fumbles lost	5.4-1
The quarterback never dumped	6.0-1
Shutting out an opponent	11.0-1
No safetys	14.0-1
No interceptions	18.0-1
No turnovers to opponent	70.0-1
No penalties	220.0-1
Never punting	280.0-1
Having undefeated season	235.0-1

Source: Ibid.

WHAT ARE THE ODDS ON WINNING IN MAJOR COLLEGE FOOTBALL (N.C.A.A. DIVISION)?

While you're less likely to see a shutout in college football than in the pros, if your team is good its odds of having an undefeated season are far greater than in pro ball.

The Odds: against (unless noted)

Shutting out opponent	9-1
Undefeated season	34-1
Major game ending in a tie	48-1
Winning by 6 points or less	7-3
Winning by 14 points or less	Even

Source: Ibid.

WHAT ARE THE ODDS ON YOUR HOME TEAM WINNING AND BECOMING CHAMPIONS IN PRO BASKETBALL?

Teams playing at home enjoy a major edge in this sport.

The Odds:

Home team winning	**7-3** for
A tie for the division championship	**30-1** against
Playoff lasting 4 games	**11-1** against
Playoff lasting 7 games	**5-3** against

Source: Ibid.

WHAT ARE THE ODDS OF WINNING BY CERTAIN POINT SPREADS IN PRO AND COLLEGE BASKETBALL?

For both pro and college basketball, very close games are the exception.

The Odds: N.B.A.

1 or 2 points	**7.3-1** against
5 or more points	**2.0-1** against
10 or more points	**5.0-4** against

The Odds: College game won by

1 or 2 points	**5-1** against
5 or more points	**2-1** against
10 or more points	Even

Source: Ibid.

WHAT ARE THE ODDS ON WINNING THE GAME IN THE CLOSING MINUTES IN PRO BASKETBALL?

About one out of six games is won in the final quarter while only four in 100 go into overtime.

The Odds:		
	Going into overtime	**23-1** against
	Two or more over	**120-1** against
	Coming from behind to win in the fourth quarter	**11-2** against

Source: Ibid.

WHAT ARE THE ODDS ON A ROOKIE EVER PLAYING IN THE MAJORS?

One out of 21 rookies will play in the majors.

The Odds: **20-1** against

Source: Ibid.

WHAT ARE THE ODDS ON A PLAYER BEING VOTED INTO THE BASEBALL HALL OF FAME?

Only one man in 167 starting line-ups will ever make it to Cooperstown.

The Odds: **1,500-1** against

Source: Ibid.

WHAT ARE THE ODDS OF YOUR HOME TEAM WINNING?

The home team has a decided edge, notably when compared with pro football.

The Odds: **1.5-1** for

Source: Ibid.

WHAT ARE THE ODDS ON WINNING BOTH GAMES OF A DOUBLE HEADER?

The team that takes the first game has a slight edge.

The Odds: **9-8** for

Source: Ibid.

WHAT ARE THE ODDS ON HOME RUNS?

Most games feature at least one homer but the odds against five or more in a single game are awesome. The odds refer to both teams.

The Odds:	1 homerun	**7-4** for
	3 or more	**7-1** against
	5 or more	**100-1** against

Source: Ibid.

WHAT ARE THE ODDS ON A CLOSE BASEBALL GAME?

A surprisingly large percentage of games go right down to the wire.

The Odds:	Game decided by one run	**5.0-2** against
	Last inning winning run	**4.3-1** against
	Going into extra innings	**10.0-1** against

Source: Ibid.

WHAT ARE THE ODDS ON HAVING NO BASES STOLEN OR ERRORS IN A GIVEN GAME?

You're far more apt to see an error free game than one in which no bases are snitched.

The Odds:	Error free game	**9-2** against
	No stolen bases	**9-5** against

Source: Ibid.

WHAT ARE THE ODDS OF SEEING A GREAT BASEBALL MOMENT?

As shown by the odds below the National League's desginated hitter rule helps. If you've seen a triple play or a no-hitter you're part of a select circle.

The Odds:	against	
	A triple play in a game	1,400-1
	A pinch hit home run:	
	National League	25-1
	American League	60-1
	A shut out	11-2
	Starter pitching a no-hitter	1,300-1
	Starter pitching a perfect game	260,000-1

Source: Ibid.

WHAT ARE THE ODDS ON BEING A 20 GAME WINNER?

We computed the odds on the basis of those who started 30 or more games.

The Odds: **5-1** against

Source: Ibid.

WHAT ARE THE ODDS ON PENNANT WINNERS AND THE DURATION OF THE WORLD SERIES?

Prior pennant winners start as better than two to one favorites to repeat. No matter who wins, a four game Series is scarce.

The Odds: against

Repeating as pennant winner	9.0-5
Repeating as World Series winner	7.0-2
A four game World Series	5.0-1
A seven game World Series	5.0-3

Source: Ibid.

WHAT ARE THE ODDS ON YOUR TEAM WINNING THE STANLEY CUP IN HOW MANY GAMES?

Unlike most championship contests, the Stanley Cup playoffs seldom last the full seven games.

The Odds:

Repeating as Stanley Cup winners	**8-3** against
4 game Stanley Cup	**5-2** against
7 game Stanley Cup	**3-1** against

Source: Ibid.

WHAT ARE THE ODDS ON YOUR TEAM WINNING AT HOME?

Hockey teams win about twice as often at home provided that you exclude the puckish tie games which help the odds of neither side.

The Odds: Home team winning   **11-6** for

Source: Ibid.

WHAT ARE THE ODDS ON COMING FROM BEHIND IN PRO HOCKEY?

With rare exception the team in the lead can protect its edge during the final minutes of play.

The Odds:		
	Winning in the last 5 minutes	**20-1** against
	Winning in the last 2 minutes	**40-1** against

Source: Ibid.

WHAT ARE THE ODDS WHEN PENALTIES ARE IN FORCE IN PRO HOCKEY?

As shown below the odds of scoring when short-handed is less of a long shot than scoring on a penalty shot.

The Odds:		
Scoring when short-handed	**35-1** against	
Being awarded penalty shot	**60-1** against	
Scoring on penalty shot	**140-1** against	

Source: Ibid.

WHAT ARE THE ODDS ON YOUR PARACHUTE MALFUNCTIONING DURING A JUMP?

According to Paraflite, Inc. you're six times better off with a square chute instead of a round one. A malfunction can mean one of two things: the chute doesn't open at all (rare) or the parachute doesn't open properly, in which case the jumper heads for earth a lot faster than he'd like to. When this happens there is always the back-up chute.

The Odds: against	
Square chute malfunction	3,000-1
Round chute malfunction	500-1
Square chute and square back-up malfunction	9,000,000-1
Square chute and round back malfunction or vice versa	1,500,000-1
Round chute and round back-up malfunction	250,000-1

Source: Data from Paraflite, Inc.

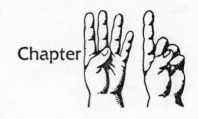

ADAM, EVE AND SOMEBODY ELSE?
Sex

Recently one of the major publishers of "skin" magazines lamented that the entire category is declining in circulation because "society has outpaced what we can produce." Is this true? In some ways yes, in some ways no and, most importantly, in many ways we simply don't know.

It's not that the world has lacked sex survey books in recent years. It sometimes seems they're going to overtake cookbooks as a leading category. What we do lack are the massive, costly studies required to adequately represent the population at large. The result is that most of the facts available are based on a certain type of respondent (readers of a magazine, attendees at a clinic, those notably interested in the women's movement, et al. Thus they don't reflect us all. Here's a case where the silent majority really is under represented!

What's more, the topic is not easy to research. When it comes to asking people what they do behind closed bedroom doors (or even more pertinently, elsewhere); many choose to remain among the silent majority.

That said, there's no question that society has become more liberated both in what we're prepared to discuss and what we're prepared to consider.

The reasons for this change are many. The new permissiveness has undoubtedly played a role, one that the mass media have been quick to exploit. The emancipation of women both sexually (with birth control) and economically is a major cause. The sexual revolution is directly related to income and education (in eight years female college enrollment has tripled to over 1,300,000). The higher these factors are the greater the incidence of sexual relations and the greater the probability of fulfillment with each one.

Withall we're hardly a swinging society. Two out of three singles have never had intercourse more than once a week, and up to the age of 25 the number of different annual sex partners is hardly a crowd: two for both single men and single women.

There are some surprises. Two out of three single women don't usually climax. You're tempted to ask why they take the risks in terms of disease and pregnancy. As we shall see, they're very high.

And then there are gay life styles about which there has been an enormous proliferation of books, films and just plain talk. Again it's a difficult area to research despite the increasing number of studies by homophile and independent research groups. As with the other sex surveys, all the data available is based on scant evidence. Part of the difficulty may well be the result of trying to determine profiles of gays in relation to heterosexual studies—a problem which most of the investigators, from Kinsey on, have found problematic. Pregnant in that approach may be value judgments—whether intended or not—which simply reinforce myths and stereotypes.

We have turned to Alan P. Bell and Martin S. Weinberg's *Homosexualities: A Study of Diversity Among Men and Women* to relieve part of the problem. Here was a study less interested in comparing homosexual and heterosexual life patterns, and centered its approach to the diversity of social and sexual profiles within a gay community. Thus, instead of sustaining myths and stereotypes, it presents data on variation within homosexual culture, and it is the odds *within* that culture we present.

In *Adam and Eve and Somebody Else* we try to give the odds of what is going on behind closed doors.

AMONG SINGLE WOMEN WHAT ARE THE ODDS ON MAXIMUM FREQUENCY OF SEX RELATED TO EDUCATION LEVEL?

Frequency of sex and higher learning go hand in hand. For example among post graduate women 37% have at one time or another had sex five or more times in one week versus 16% of those who at most finished high school.

The Odds: against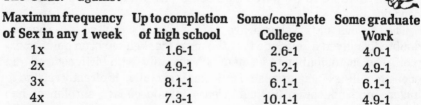

Maximum frequency of Sex in any 1 week	Up to completion of high school	Some/complete College	Some graduate Work
1x	1.6-1	2.6-1	4.0-1
2x	4.9-1	5.2-1	4.9-1
3x	8.1-1	6.1-1	6.1-1
4x	7.3-1	10.1-1	4.9-1
5x or more	5.2-1	3.8-1	1.7-1

Source: Alfred Kinsey, *Sexual Behavior in the Human Male*, Philadelphia, Pa., W.B. Saunders Co., 1948, p. 236.

WHAT ARE THE ODDS ON A FEMALE LOSING HER VIRGINITY AT CERTAIN AGES COMPARED WITH HER EDUCATIONAL LEVEL?

Quite clearly the less a woman wants to hit the books in school the more she's apt to turn to sex as a substitute. Almost nine out of 10 women who only finish grade school are no longer virgins by the age of 18, whereas among college graduates three out of four were still virgins at this age.

The Odds: against unless noted

Age at loss	Grade school	Graduate of high school	Some college	College or advance degree
15 & under	1.4-1 for	4.0-1	10.1-1	13.3-1
16-17	2.6-1	Even	2.1-1	4.3-1
18-19	9.0-1	3.3-1	1.3-1	2.2-1
20-21	99.0-1	19.0-1	7.3-1	2.7-1
22 +	49.0-1	32.3-1	10.0-1	5.2-1

Source: Carol Tavris and Susan Sadd, *The Redbook Report on Female Sexuality*, New York, Dell Publishing Company, Inc., © 1975, 1977 by the Redbook Publishing Company, p. 73.

WHAT ARE THE ODDS ON A FEMALE LOSING HER VIRGINITY, ACCORDING TO AGE AND RACE?

By age 19 half of all white women and over eight out of 10 black women have lost their virginity. Comparative figures for 1971 (not shown here) reveal that more women are having sex and doing so at an earlier age.

The Odds: against unless noted

Age	All women	White	Black
15	4.6-1	6.2-1	1.6-1
16	2.9-1	3.4-1	1.1-1 for
17	1.4-1	1.8-1	2.2-1 for
18	1.2-1	1.3-1	2.9-1 for
19	1.2-1 for	Even	5.1-1 for
All 15-19	1.9-1	2.2-1	1.7-1 for

Source: Melvin Zelnik and John F. Kanter, "Sexual and Contraceptive Experience of Young Unmarried Women in the United States," *Family Planning Perspective*, Vol. 9, No. 2, 1977.

WHAT ARE THE ODDS ON WHERE WOMEN WILL BE WHEN THEY LOST THEIR VIRGINITY?

When it comes to a woman's first sexual intercourse, the answer to the age-old question "your place or mine" is quite clearly "yours." Four out of 10 women aged 15 to 19 lose their virginity in the man's home. The closest runner-up is the home of a relative or friend. This accounts for one of five first experiences. There isn't much difference between blacks and whites on the above figures. However, blacks went to motels and/or hotels far more.

The Odds: against

	All 15-19 year old single women	Whites	Blacks
Woman's home	6.1-1	6.7-1	4.8-1
Partner's home	2.4-1	2.3-1	2.6-1
Relative/friend	4.7-1	5.0-1	4.8-1
Motel/hotel	19.2-1	45.4-1	7.4-1
Automobile	10.5-1	9.0-1	20.4-1
Other	16.6-1	13.0-1	71.4-1

Source: *Family Planning Perspectives*, Vol 9, No. 2, March/April 1977.

IF YOU'RE SINGLE WHAT ARE THE ODDS THAT YOU'LL SLEEP WITH SOMEONE TONIGHT?

Unfortunately we were only able to uncover data on this question for those aged 18 to 24. Seventy-five percent of all single men in the group had premarital sex in the past year with a medium frequency of 37 times. Two thirds of all single women have had sex in the past year. Among this group the median frequency is far higher than that of single men: just over once a week. What accounts for the difference between the two groups of singles? Married men.

The Odds: Men   **11.8-1** against   Women   **9.0-1** against

Source: Morton Hunt, *Sexual Behavior in the 1970's*, New York, N.Y., Dell Publishing Co., 1974, 1974, p. 167, 168.

IF YOU'RE SINGLE WHAT ARE THE ODDS THAT YOU'LL SLEEP WITH SOMEONE NEW TONIGHT?

You'll have to compute these odds based on your own sexual experience. Both single men and single women under the age of 25 have an average of two sexual partners a year. Between the ages of 25 and 34 that rises to four for men and three for women. But we don't know how many of these are new lovers or how many are carryovers from previous years. The odds below assume a fresh start, as it were, each year.

The Odds:	Men 25	**181.5-1** against
	Men 25 to 34	**90.2-1** against
The Odds:	Women 25 years	**181.5-1** against
	Women 25 to 34	**120.6-1** against

Source: Ibid.

WHAT ARE THE ODDS THAT YOU'LL HAVE SEX X NUMBER OF TIMES PER WEEK IF YOU'RE SINGLE?

Almost one out of three singles (30%) has never had sex more than once a week. The vast majority (69%) have never had sex more than four times per week. At the other end of the activity scale are 13% who have had sex at least eight times a week. Included in this group are a hardy band (4%) who've had sex 20 or more times in one week. So much for the bedroom athletes!

The Odds: against	Maximum frequency of sex in any one week	
	1x	2.3-1
	2x	5.3-1
	3x	6.7-1
	4x	9.0-1
	5x	15.7-1
	6x	24.0-1
	7x	13.3-1
	8 + x	6.7-1

Source: Kinsey, p. 335.

FOR WOMEN, WHAT ARE THE ODDS ON EXPERIENCING AN ORGASM?

For married women the odds are better than even on any given instance, while for single women intercourse would appear to be substantially less rewarding: over two out of three don't normally reach a climax. The figures cited here are generally supported by other studies.

The Odds: against	Married women	Single women
All the time	5.7-1	13.9-1
Most of the time	1.1-1	3.6-1
Sometimes	4.2-1	1.7-1
Once in a while	8.4-1	——
Never	13.3-1	1.9-1

Source: *Redbook Report on Female Sexuality*

WHAT ARE THE ODDS THAT YOU'LL CONTRACT A VENEREAL DISEASE DURING THE COURSE OF A YEAR?

For the latest full reported year there were 597,000 cases of male gonorrhea and 403,000 cases among women. In addition there were 15,000 reported cases of syphillis among men, 5,000 among women. Many experts think that at least two cases go unreported for every one recorded.

The Odds:	Based on reported cases	Based on estimated total cases	
Men over 14	**130-1** against	**42.8-1** against	☞
Women over 14	**217-1** against	**71.7-1** against	

Source: Department of Health, Education, and Welfare, *Sexually Transmitted Disease Statistics Letter*, No. 127, 1978.

WHAT ARE THE ODDS ON MARRIED WOMEN USING OR NOT USING VARIOUS MEANS OF CONTRACEPTION, BY RACE AND AGE?

Seven out of 10 white wives and six out of 10 black wives use some form of contraception. In both groups the pill is the preferred form of birth control. For sexually active young single women (15-19) Family Planning Perspectives reported that 63% used some form of contraceptive at last intercourse; the pill was far and away the preferred means for 40% of whites and 60% of blacks.

The Odds: against unless noted

	Whites			
	15-24	**25-34**	**35-44**	
Contraceptive used	2.2-1 for	2.6-1 for	1.9-1 for	☞
Sterilized	27.6-1	3.8-1	2.3-1	
Pill	1.3-1	3.3-1	11.8-1	
IUD	14.8-1	13.1-1	20.7-1	
Diaphragm	34.7-1	30.2-1	32.3-1	
Condom	18.6-1	12.1-1	10.6-1	
Foam	33.5-1	32.3-1	34.7-1	
Rhythm	40.7-1	29.3-1	22.2-1	
All other	51.6-1	27.5-1	22.2-1	

	Blacks			
	15-24	**25-34**	**35-44**	
Contraceptive used	1.4-1 for	1.6-1 for	1.1-1 for	☞
Sterilized	24.0-1	9.3-1	3.4-1	
Pill	1.8-1	2.8-1	12.0-1	
IUD	16.2-1	13.1-1	19.0-1	
Diaphragm	499.0-1	57.8-1	33.5-1	
Condom	28.4-1	17.2-1	23.4-1	
Foam	54.5-1	22.2-1	21.2-1	
Rhythm	28.4-1	249.0-1	70.4-1	
All other	23.4-1	12.5-1	19.0-1	

Source: Department of Commerce, Bureau of the Census, *Statistical Abstract of the United States*, 1978.

WHAT ARE THE ODDS BY AGE THAT A MAN WILL BECOME TEMPORARILY OR PERMANENTLY IMPOTENT?

According to studies reported in Medical Aspects of Human Sexuality (January, 1976), "an estimated one half of the male population has experienced occasional transient episodes of problems in attaining and/or sustaining an erection." So when it comes to temporary problems a man's odds are even.

The Odds: against unless noted

Under 30	124.0-1
By age 40	51.6-1
By age 50	13.9-1
By age 55	13.9-1
By age 60	4.4-1
By age 65	3.0-1
By age 70	2.7-1
By age 75	1.2-1 for
By age 80	3.0-1 for

Source: Kinsey, p. 236

WHAT ARE THE ODDS THAT YOU'LL HAVE A HOMOSEXUAL OR LESBIAN RELATIONSHIP?

According to Morton Hunt's research for his book Sexual Behavior in the 1970's, about 13% of men and 11% of women have had an overt experience with one or more partners of their own sex. Only 7% of all males and 3% of all females have had a homosexual experience for more than three years of their lives.

The Odds: Men   **7.7-1** against

             Women **8.1-1** against

Source: Hunt, pp. 305-317.

WHAT ARE THE ODDS THAT BOYS WILL BE BOYS?

According to the Sex Information and Education Council of the U.S., various studies show that 37% of males have at least one homosexual experience which leads to orgasm, and 8% of males are homosexuals throughout their lifetime. The Gay Activist Front disputes this latter figure, maintaining it should be closer to 10%.

To compute our odds we used the more conservative figure to determine who aren't boys for a lifetime.

The Odds: **39.0-1** for

Source: Data from the Sex Information and Education Council of the U.S.

WHAT ARE THE ODDS OF HOMOSEXUAL MEN AND WOMEN FEELING EXCLUSIVELY HOMOSEXUAL?

Sexual identity and gender orientation have never, as Kinsey taught us many years ago, been clear cut issues. White men and black women seem to have made stronger commitments to their sexual orientation.

The Odds:	White male	**1.2** for
	Black male	**1.4** against
	White female	**1.2** against
	Black female	**1.1** for

Source: Bell, Alan P. and Weinberg, Martin S., Institute for Sex Research, *Homosexualities*, New York, Simon and Schuster, 1978.

WHAT ARE THE ODDS THAT PARENTS KNOW THEIR CHILDREN ARE HOMOSEXUAL?

Despite the increasing trend to "come out of the closet," declaration of sexual preference to family members is still a sensitive issue. Mother is more likely to know than father, and blacks seem more likely to reveal that they're gay than whites.

The Odds: against unless noted

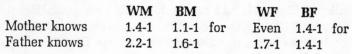

	WM	BM		WF	BF	
Mother knows	1.4-1	1.1-1	for	Even	1.4-1	for
Father knows	2.2-1	1.6-1		1.7-1	1.4-1	

Source: Ibid.

WHAT ARE THE ODDS THAT A HOMOSEXUAL HAS BEEN MARRIED?

Men are at least twice as likely not to have been previously married.

The Odds: against

	White male	4.0-1
	Black male	6.7-1
	White female	1.9-1
	Black female	1.1-1

Source: Ibid.

WM = White males   BM = Black males   WF = White females   BF = Black females

WHAT ARE THE ODDS OF REVEALING HOMOSEXUAL ORIENTATION AT WORK?

Full disclosure of being gay is not likely either to the employer or to fellow employees.

The Odds:

That the employer doesn't know

White male	1.1-1 for
Black male	1.9-1 for
White female	2.1-1 for
Black female	1.5-1 for

That most of the fellow workers know

White male	2.4-1 against
Black male	3.2-1 against
White female	5.7-1 against
Black female	6.7-1 against

Source: Ibid.

WHAT ARE THE ODDS THAT HOMOSEXUALITY WAS IN SOME WAY RESPONSIBLE FOR THE ENDING OF THE MARRIAGE?

The results are decisively divided on racial lines:

The Odds:

White male	1.2-1 for	White female	1.2-1 for
Black male	3.4-1 against	Black famale	1.4-1 against

Source: Ibid.

WHAT ARE THE ODDS OF HOMOSEXUAL MEN AND WOMEN HAVING MADE AN ATTEMPT TO GO "STRAIGHT"?

With the exception of black males, the odds favor that most will have tried.

The Odds:

White male	**1.7-1** for	White female	**3.0-1** for
Black male	**1.6-1** against	Black female	**1.4-1** for

Source: Ibid.

WM = White males   BM = Black males   WF = White females   BF = Black females

WHAT ARE THE ODDS ON THE NUMBER OF TIMES GAY MEN AND WOMEN WILL GO OUT TO BARS?

Gay bars—like the local neighborhood pub—are used as an important place for meeting and socializing with friends. And like their heterosexual counterparts—the singles bars—they are the "cruising" grounds for sexual partners. This latter fact may well account for the frequency with which they are attended, especially by men.

The Odds: against	WM	BM	WF	BF	
Once a month	8.1-1	19.0-1	5.3-1	8.1-1	
2-3x a month	6.7-1	3.2-1	6.1-1	4.0-1	
Once a week	4.0-1	4.9-1	4.9-1	3.0-1	
2-3x a week	2.5-1	1.1-1	8.1-1	11.5-1	

Source: Ibid.

WHAT ARE THE ODDS THAT GAYS CONSIDER THEMSELVES REPUBLICAN, DEMOCRAT OR INDEPENDENT?

Despite the fact that the data base for this question was admittedly small, if there is any political food for thought it is that Republican platforms have not proven appealing to members of the gay community.

The Odds: against unless noted

	WM	BM	WF	BF	
Democrat	1.8-1	Even	1.1-1	1.1-1 for	
Republican	4.9-1	32.3-1	9.0-1	0	
Independent	1.4-1	1.2-1	2.2-1	1.4-1	

Source: Ibid.

WHAT ARE THE ODDS OF HOMOSEXUAL MEN AND WOMEN CONTRACTING A VENERAL DISEASE?

About two thirds of homosexual men have contracted veneral disease at least once. White lesbians almost never contract veneral diseases as the phenomenal odds indicate.

The Odds:

White male	**1.6-1** for	
Black male	**2.9-1** for	
White female	**999.0-1** against	
Black female	**49.0-1** against	

Source: Ibid.

WM = White males   BM = Black males   WF = White females   BF = Black females

WHAT ARE THE ODDS THAT GAY MEN AND WOMEN WILL HAVE SEX X NUMBER OF TIMES PER WEEK?

Just short of seven out of 10 white males (69%) have sex more than once a week, while among black males, the figure is slightly more than four out of five (82%). Although lesbian women indulge in sex less frequently than men, the two-to-three-time frequency is most probable, especially if we were to combine all the categories into a single number.

The Odds: against unless noted

	WM	BM	WF	BF
Less than once a week	2.0-1	4.6-1	Even	1.9-1
1x	3.5-1	4.9-1	4.0-1	10.1-1
2-3x	2.3-1	1.4-1	4.3-1	1.6-1
4-6x	6.7-1	5.3-1	10.1-1	8.1-1
7 +	24.0-1	13.3-1	24.0-1	15.7-1

Source: Ibid.

WHAT ARE THE ODDS ON THE NUMBER OF SEXUAL PARTNERS GAY MEN AND WOMEN WILL HAVE?

There is no question about it: especially among gay males, sexual contact is frequent and involves an extraordinary number (hundreds) of partners. Whatever the reasons, the image of the homosexual male as promiscuous is likely to be a sure bet. Lesbians, on the other hand, favor long term relationships, sexual fidelity, and demonstrate bonding patterns that are far closer to traditional marriage.

The Odds: against:

No. of Partners	WM	BM	WF	BF
1	0	0	32.3-1	19.0-1
2	0	0	10.1-1	19.0-1
3-4	99.0-1	49.0-1	5.7-1	6.1-1
5-9	49.0-1	24.0-1	2.2-1	2.3-1
10-14	32.3-1	19.0-1	5.3-1	10.1-1
15-24	32.3-1	15.7-1	9.0-1	8.1-1
24-49	11.5-1	15.7-1	11.5-1	8.1-1
50-99	10.1-1	4.6-1	19.0-1	11.5-1
100-249	5.7-1	5.7-1	99.0-1	49.0-1
250-499	4.9-1	8.1-1	99.0-1	49.0-1
500-999	5.7-1	6.1-1	0	0
1000 or more	2.6-1	4.3-1	0	0

Source: Ibid.

WM = White males   BM = Black males   WF = White females   BF = Black females

WHAT ARE THE ODDS THAT GAY MEN AND WOMEN WILL SLEEP WITH SOMEONE NEW TONIGHT?

We divided the answers to this question into two parts: the proportion of one-night stands to all sexual encounters (in which category the partner may have been known socially) and the proportion of partners who are strangers (and who may eventually become repeats). Especially when one considers the sheer number of partners that the last set of odds describes, the figures below confirm that for white male homosexuals, coming and going predominates. Lesbians (who have far fewer partners than men) lean heavily away from having sex with someone they don't know.

The Odds: against unless noted

Proportion of one-night-stand partners	WM	BM	WF	BF
None	99.0-1	24.0-1	1.6-1	1.4-1
Half or less	2.5-1	1.4-1 for	Even	1.2-1 for
More than half	2.3-1 for	1.6-1	7.3-1	19.0-1

Proportion of partners who are strangers				
None	99.0-1	19.0-1	1.6-1 for	1.3-1 for
Half or less	4.0-1	1.3-1	2.1-1	1.6-1
More than half	3.8-1	Even	15.7-1	15.7-1

Source: Ibid.

WHAT ARE THE ODDS ON HOW LONG THE FIRST AFFAIR WILL LAST?

Getting through the first year seems difficult enough, but after three years, first affairs are a good bet to go bad.

The Odds: against

	WM	BN	WF	BF
3 months or less	5.3-1	7.3-1	10.0-1	32.3-1
4-11 months	3.2-1	3.5-1	4.6-1	3.5-1
1-3 years	1.6-1	1.1-1	1.4-1	1.1-1
4-5 years	9.0-1	8.1-1	4.9-1	3.8-1
More than 5 years	7.3-1	11.5-1	6.1-1	15.7-1

Source: Ibid.

WM = White males   BM = Black males   WF = White females   BF = Black females

WHAT ARE THE ODDS ON THE NUMBER OF YEAR DIFFERENCE IN AGE IN AN AFFAIR?

The best overall chance for compatability seems to demand a partner who is one to five years older or younger. But one thing's for sure: if you and your lover are the same age, the odds are decidedly against the affair working.

The Odds: against

Age difference	WM	BM	WF	BF	
0	19.0-1	9.0-1	9.0-1	32.3-1	☞
1-2 years	3.5-1	3.0-1	3.3-1	2.2-1	
3-5 years	3.8-1	3.0-1	2.0-1	4.3-1	
6-10 years	3.3-1	4.0-1	4.3-1	2.6-1	
10 + years	2.6-1	4.0-1	5.3-1	4.3-1	

Source: Ibid.

WHAT ARE THE ODDS THAT HOMOSEXUAL COUPLES WILL COMBINE INCOMES?

At least economically, men are likely to keep their independence, whereas women tend to reinforce their bonds with joint-accounting.

The Odds: White male   **1.1-1** against   ☞
White male
Black male   **1.6-1** against
White female   **1.3-1** for
Black female   **1.2-1** for

Source: Ibid.

WHAT ARE THE ODDS THAT THE HOUSEWORK IS SHARED?

Heterosexuals may find a lesson here: love and long term relationships are likely to demand equal time with the vacuum cleaner and washing the dishes.

The Odds: White male   **1.6-1** for   ☞
Black male   **2.2-1** for
White female   **1.4-1** for
Black female   **2.5-1** for

Source: Ibid.

WM = White males   BM = Black males   WF = White females   BF = Black females

Chapter

TYING AND UNTYING THE KNOT
Marriage and Divorce

Of all Americans over the age of 50 only 6% of men and 4% of women have never wed. But the part of the ceremony that goes "till death do us part" could well be amended to "till death do us part—or the laywers." Currently 38% of first marriages end in divorce; that's almost four out of 10. While the vast majority of those who divorce remarry (four out of five), the odds, as you'll see, on a remarriage are even less favorable.

Why then even bother to marry at all? It's a question that many men and women seem to be taking seriously. In 1978 two million two hundred thousand people were living together as unmarried couples, a figure that's increased eightfold in the last eight years. Furthermore in just over one out of five of these households there's at least one child present either from the current liaison or from a previous one. So the old adage "not in front of children" is somewhat out of date.

Children still play a big part in marriages: they cause them. In spite of the widespread acceptance of both the pill and abortion, latest reported figures (1972-1976) show that in just under one in four marriages with a bride aged 14 to 24 she was either pregnant or had already had a child.

These marriages are concentrated among couples in ethnic minorities with far less education and lower incomes than the population at large. And these marriages are among those which are most apt to break up by separation, divorce or, quite simply, abandonment. Study after study shows that the biggest single cause of family disintegration isn't sex but quarrels over money. Also, in a large number of these marriages one or both parties came from single parent homes. It's a classic vicious spiral, something we'll encounter often in this book.

Divorce seems to have become as American as apple pie. Whereas one in four marriages broke up in 1970, the current ratio is one in three. So the odds on a marriage succeeding are only two to one, for.

The growth in the divorce rate can be traced to many causes. Two which have contributed significantly are changed public attitudes and increased female financial independence. The former is the continuation of a long term trend. In most of America divorce has become increasingly socially acceptable.

The higher number of working wives is a different matter. This has grown at such a startling rate that today very nearly one out of two wives works (versus 40% as late as 1970). Sociologists across the land are busy explaining this phenomenon. There's general agreement that it has contributed to an increased sense of female independence. As odds in the chapter show it has certainly contributed to sexual independence. Over half of all 35 to 39-year-old full time working wives have had at least one affair.

For a mother, a new found sense of independence is one reality, but the consequences of divorce are another hard one. Long term alimony awards are decreasing in favor of short-term "rehabilitation payments" to help a woman retool her skills. Furthermore, child support payments seldom reflect needs: latest available figures show that on the average fathers only pay a maximum of 16% of their gross income. Even if a woman does go out to work, on the average she'll only earn 60% of her husband's wage. And who's going to look after the kids? Finally, an appallingly large number of court agreed awards turn out to be meaningless scraps of paper. Vast numbers of husbands and fathers never or seldom pay, while only a minority scrupulously honor their obligations.

In *Tying and Untying the Knot* we set out to examine all aspects of one of life's biggest commitments, marriage, to analyze the causes and consequences of what is often one of life's bigger disturbances – the all-too-frequent subsequent divorce – and to highlight some of the results.

WHAT ARE THE ODDS THAT THE HONEYMOON IS OVER?

According to the U.S. Census Bureau if you total all marriages in the United States their average length is 23.6 years. Industry estimates indicate an average honeymoon length of 21 days.

The Odds: **4,092-1 for**

Source: Department of Commerce, Bureau of the Census.

WHAT ARE THE ODDS THAT YOU'LL MARRY SOMEONE WHO'S OLDER/YOUNGER THAN YOU ARE?

While one out of two husbands is up to four years older than their spouse, more husbands are younger than their wives (14.4%) than those that are the same age (12.1%).

The Odds: against

Husband's age	
10 + years older	13.8-1
5-9 years older	5.1-1
3-4 years older	4.9-1
1-2 years older	3.8-1
Same age	9.0-1
Younger	6.9-1

Source: Department of Commerce, Bureau of the Census, P-23, No. 77, 1977, p. 5.

WHAT ARE THE ODDS THAT A WOMAN WILL BE A VIRGIN WHEN SHE MARRIES?

The chastity odds are slim and getting smaller all the time. Ninety-one percent of women who married before 1964 were virgins; by 1973 only seven out of every 100 brides were chaste.

The Odds: **13.3-1** against

Source: Carol Tavris and Susan Sadd, *The Redbook Report on Female Sexuality,* New York, Dell Publishing Company, Inc., ©1975, 1977 by the Redbook Publishing Company.

WHAT ARE THE ODDS THAT A FEMALE WILL MARRY THE FIRST PERSON SHE SLEEPS WITH?

The only data we could turn up on that was the Kinsey report of the late 1940's. At that time 46% of married women had only had premarital sex with the man they married.

The Odds: **1.2-1** against

Source: Alfred Kinsey, *Sexual Behavior in the Human Male,* Philadelphia, Pa., W.B. Saunders Co., 1948.

WHAT ARE THE ODDS THAT THE WIFE WILL BE PREGNANT WHEN THE MARRIAGE TAKES PLACE?

The Census Bureau has analyzed births relative to marital dates. These imply that in one out of five cases the bride is already pregnant.

The Odds: **4-1** against

Source: Department of Commerce, Bureau of the Census.

WHAT ARE THE ODDS ON HOW OFTEN SEX TAKES PLACE EACH MONTH IN A MARRIAGE?

Just over 50% of couples have sex from six to 15 times per month while one in four only have sex weekly or less frequently. One couple in 12 have sex 20 or more times each month. Only 11% of women in this latter group feel this is too much sex while 5% actually feel they are still getting too little.

The Odds: against

Monthly Number of Sexual Encounters	
0	49.0-1
1-5	2.8-1
6-10	2.1-1
11-15	3.8-1
16-20	8.1-1
20 +	11.5-1

Source: *The Redbook Report on Female Sexuality*

WHAT ARE THE ODDS THAT THE WIFE IS SEXUALLY UNFULFILLED IN THE MARRIAGE?

According to a *Redbook* report 57.6% of women feel the frequency of intercourse in their marriages is about right while almost four in 10 (38.1%) wish they had more frequent sex.

The Odds: **Frequency of sex**

About right	**1.4-1** for	
Too frequent	**22.2-1** against	
Too infrequent	**1.6-1** against	

Source: *The Redbook Report on Female Sexuality*

WHAT ARE THE ODDS THAT YOUR HUSBAND/WIFE HAS EVER HAD AN AFFAIR?

Among all married women 29% have had an affair. Half of these have only had one such extramarital experience while one out of 20 has had over 10 partners. If an affair does take place it's not apt to be a single chance encounter. Only 6.1% of affairs are one time sexual encounters. In three out of four cases lovemaking took place over 10 times.

Of married white men, including those now divorced, one out of two has had at least one affair.

The Odds: against

All married women
Age

25	4.0-1
25-29	2.4-1
30-34	2.3-1
35-39	1.6-1
40 +	1.5-1

Source: *The Redbook Report on Female Sexuality*

Ever married white men
Age

25	2.1-1
25-34	1.4-1
35-44	1.1-1
45-54	1.6-1
55 +	1.3-1

Source: Morton Hunt, *Sexual Behavior in the 1970's*, New York, Dell Publishing Co., Inc. ©Morton Hunt, 1974.

WHAT ARE THE ODDS THAT A WIFE WILL HAVE EXTRAMARITAL AFFAIRS BASED ON THE AGE WHEN SHE FIRST HAD SEX?

There is a direct correlation between age at loss of virginity and the odds on cheating on a husband later. Among women who were over 21 when they first had premarital sex only 16% subsequently cheated on their husbands. Yet of all women who were 15 or younger when they lost their virginity almost 50% (48 out of 100) subsequently cheat on their husbands.

The Odds: Women over 21 when
          first had premarital sex   **5.2-1 against**
          Women 15 or under when
          first had premarital sex   **Even**

Source: *The Redbook Report on Female Sexuality*

WHAT ARE THE ODDS BY AGE AND EMPLOYMENT STATUS ON MARRIED WOMEN HAVING AN AFFAIR?

By the time married women reach the age of 40, four out of 10 have had at least one extramarital affair. In all cases the figures are highest for working wives. Indeed, as shown in the table, by the time a working wife is in her late 30's the odds are better than even that she has had at least one man on the side.

The Odds: against unless noted

Age	Full time employed wives	Housewives
Under 25	3.8-1	4.9-1
25-34	1.5-1	3.3-1
35-39	1.1-1 for	3.2-1
40	1.1-1	2.0-1

Source: *The Redbook Report on Female Sexuality*

WHAT ARE THE ODDS AMONG MARRIED WOMEN WHO HAVE HAD AN AFFAIR ON THE NUMBER OF AFFAIRS THEY'LL HAVE?

Half of all women who have an extramarital fling will stop at one, while a further four out of 10 will have five or fewer affairs. Conversely one out of 20 women who cheats on her husband will do so 10 or more times.

The Odds:	1 affair	**Even**
	2-5 affairs	**1.5-1** against
	6-10 affairs	**19.0-1** against
	11 or more	**19.0-1** against

Source: *The Redbook Report on Female Sexuality*

WHAT ARE THE ODDS ON THE NUMBER OF SEXUAL ENCOUNTERS IN AN AFFAIR?

Over 40% of women have at least had sex with each man 10 times or more, while just 6% have confined themselves to a single sexual encounter. Roughly one out of four reported the number of encounters varied widely by partner.

The Odds: against

1 encounter	15.4-1
2-5 encounters	3.9-1
6-10 encounters	13.1-1
More than 10	1.5-1
Varied greatly from partner to partner	2.8-1

Source: *The Redbook Report on Female Sexuality*

WHAT ARE THE ODDS THAT YOU'LL "SWING" DURING YOUR MARRIAGE?

Four percent of married couples have swapped partners during the course of their marriage. This figure remains remarkably constant among couples of all ages. Of those that have tried mate swapping 54% have done so just once, 33% occasionally and 13% often.

The Odds: **24.0-1** against

Source: *The Redbook Report on Female Sexuality*

WHAT ARE THE ODDS THAT THE THRILL IS GONE?

In May, 1978 the Journal of Sexual Research reported the results of a probability sample analyzing middle class couples on what it called "interpersonal sexuality" versus a preference for other activities based on the number of years married. Over the course of the years, given the choice between their husband and a good book, women preferred the book (37% to 26%). While men continue to opt for sex (45%), going to sports events is a close second (41%). These are the odds on preferring sex to other activities.

The Odds: against unless noted

Number of years Married	Husbands	Wives	
0-5	1.5-1 for	2.4-1	
6-10	1.7-1 for	3.1-1	
11-15	1.8-1 for	4.0-1	
16-20	2.3-1	3.2-1	
21-25	3.2-1	5.3-1	
26-30	2.7-1	3.8-1	
31-35	3.3-1	7.7-1	
36-50	7.1-1	59-1	

Source: Jay A. Mancini and Dennis K. Orthner, "Recreational Sexuality Preferences Among Middle-Class Husbands and Wives," *Journal of Sexual Research*, Vol. 14, No. 2, 1978, pp. 96-106.

WHAT ARE THE ODDS THAT YOU'LL GET DIVORCED THIS YEAR?

In 1978, there were 1,122,000 divorces out of 47,631,000 husband and wife households.

The Odds: **41.4-1** against

Source: Monthly Vital Statistics, *Annual Summary of the United States,* —1978.

WHAT ARE THE ODDS ON A DIVORCE INVOLVING CHILDREN?

In almost seven out of 10 marriages where a divorce takes place the couple are childless.

The Odds: **1.3-1** against

Source: Department of Commerce, Bureau of the Census, *Divorce, Child Custody, and Child Support*, 1979.

WHAT ARE THE ODDS BASED ON AGE AND SEX OF YOUR FIRST MARRIAGE ENDING IN DIVORCE?

Two truths apply here. The younger you are the more apt your marriage is to ultimately end in divorce, and women of all ages divorce more than men.

The Odds: against

Year of Birth	Men	Women
1900-1904	7.1-1	7.1-1
1905-1909	6.7-1	6.7-1
1910-1914	5.8-1	6.2-1
1915-1919	5.5-1	5.8-1
1920-1924	5.0-1	5.0-1
1925-1929	4.5-1	4.2-1
1930-1934	4.2-1	3.8-1
1935-1939	3.4-1	3.2-1
1940-1944	3.2-1	2.9-1
1945-1949	2.9-1	2.6-1

Source: Department of Health, Education and Welfare, National Center for Health Statistics, *Divorces and Divorce Rates*, 1978.

WHAT ARE THE ODDS BY NUMBER OF YEARS MARRIED ON DIVORCING?

Over three out of 100 marriages fail within the first year. Is this the real danger period? As the figures below show the second and third years of marriage are the high (or low) points. In each of these about four out of 100 couples decide to call it quits.

The seven year itch is part of our mythology. There appears to be something to it. If you stay married for seven years the odds on remaining married escalate very quickly.

The Odds: against

Less than 1 year	35.4-1
1 year	29.6-1
2 years	25.5-1
3 years	25.4-1
4 years	26.8-1
5 years	28.8-1
6 years	33.4-1
7 years	35.2-1
8 years	40.5-1
9 years	45.7-1
10-14 years	59.2-1
15-19 years	87.7-1
20-24 years	128.2-1
25-29 years	200.0-1
30 + years	625.0-1

Source: Ibid.

WHAT ARE THE ODDS ON A WOMAN BEING AWARDED ALIMONY?

The odds are non-existent if you happen to live in Dallas or Philadelphia. Texas and Pennsylvania laws ban alimony entirely. Only 14% of all separation and divorce agreements award alimony.

The Odds: **6.1-1** against

Source: Department of Health, Education and Welfare, National Center for Health Statistics, *Divorces and Divorce Rates*, 1978.

IF AWARDED ALIMONY WHAT ARE THE ODDS THAT IT WILL BE PAID REGULARLY?

Over two out of 10 men don't make even one payment, while over two more rarely pay and one (8%) only pays sometimes.

The Odds: **1.2-1** against

Source: Ibid.

IF ALIMONY ISN'T PAID WHAT ARE THE ODDS THAT THE WOMAN WILL INITIATE LEGAL ACTION?

Just under two out of three women take legal action to collect.

The Odds: **1.7-1** for

Source: Ibid.

IF YOU'RE A DIVORCED, SEPARATED OR SINGLE WOMAN WHAT ARE THE OVERALL ODDS OF GETTING CHILD SUPPORT?

Some 40% of mothers never receive support from the man for their children. The reasons are many: failure to establish a legal obligation (especially true with children born out of wedlock), a court's decision that the father can't afford to pay in any event, a division of property at the time the relationship was terminated, support from the woman's family, the mother's desire to have no further communication, or simple inability to find the father.

The Odds: **1.5-1** against

Source: Ibid.

IF A WOMAN IS AWARDED CHILD SUPPORT WHAT ARE THE ODDS THAT IT WILL BE PAID REGULARLY?

Sixteen percent of men don't make a single payment. Of the balance, one third miss their payments for at least a full year, while a further third pay something but only at irregular intervals.

Whether or not a father pays, how much he pays and the regularity with which he keeps up payments are all directly related to his income. Thus less than one in five fathers in the lower one fourth income group ever pays a penny, while three fourths of those in the top earning quarter make at least one payment.

The Odds: **4.5-1** against

Source: Ibid.

WHAT ARE THE ODDS ON THE AGE OF THE MOTHER AND THE NUMBER OF CHILDREN INVOLVED AT THE TIME OF DIVORCE?

About two thirds of all divorced people (64.5%) are under the age of 29. Among this group one third have no children and a further third have one child. The larger the family the less likely divorce is. The likelihood of divorce also declines as women grow older—along with their marriages.

The Odds: against unless noted

Aged 14 to 29 at divorce	1.8-1 for
No children	3.7-1
1 child	3.7-1
2 children	6.4-1

The Odds: against unless noted

3-5 children	2.3-1
Aged 30 to 39 at divorce	3.1-1
Aged 40 to 75 at divorce	7.8-1

Source: Department of Commerce, Bureau of the Census, *Current Population Reports*, Series P-20, No. 312, "Marriage, Divorce, Widowhood, and Remarriage by Family Characteristics," 1975.

WHAT ARE THE ODDS BY RACE THAT IN A ONE PARENT FAMILY MAINTAINED BY A WOMAN SHE'LL HAVE CERTAIN CHARACTERISTICS?

Black single mothers tend to be younger (57.2% are under 35) than white single mothers (47.9%). While over nine out of 10 white single mothers were once married, one third of black single mothers have never married.

The Odds: against unless noted

Marital status of mother:	Black	White
Never married	2.0-1	11.7-1
Separated	2.4-1	3.7-1
Divorced	3.3-1	1.1-1 for
Widowed	9.1-1	6.2-1
Married spouse, absent,		
except separated	2.6-1	18.2-1
Age of mother:		
Under 35 years	1.3-1 for	1.1-1
35 to 44 years	2.5-1	2.1-1
45 to 64 years	6.3-1	4.2-1
65 years and over	49.9-1	33.2-1

Source: Department of Commerce, Bureau of the Census, *Current Population Reports*, Series P-20, No. 291, "Household and Family Characteristics," 1975 and data from Current Population Survey.

WHAT ARE THE ODDS ON RECEIVING CHILD SUPPORT AND IF SO HOW MUCH?

An appalling 95% of mothers don't get a penny's worth of support from the father, and more than 80% of separated women and almost 60% of divorced mothers suffer a similar fate. Among the minority who do receive payment, on the average unmarried mothers receive only about half the amount ($1,503) received by separated ($3,178) or divorced ($2,836) women.

The Odds: against unless noted

Marital status of women	Received child support in year reported	Amounts (if received)			
		$1-999	$1000-1499	$1500-1999	$2000-2999
Never married	23.4-1	1.6-1 for	2.7-1	-	11.3-1
Currently separated	4.5-1	1.6-1	6.8-1	8.6-1	6.8-1
Currently divorced	24-1	2.2-1	4.4-1	7.8-1	6.3-1

Marital status of women	$3000-3999	$4000-4999	$5000-6999	$7000+
Never married	-	36.0-1	-	-
Currently separated	30.3-1	24.0-1	124-1	19.8-1
Currently divorced	15.0-1	36.4-1	45.8-1	139-1

Source: Department of Commerce, Bureau of the Census, *Divorce, Child Custody, and Child Support*, pp. 13, 14.

WHAT ARE THE ODDS ON A SINGLE MOTHER RECEIVING CHILD SUPPORT BY VARIOUS CHARACTERISTICS?

As the odds show educated mothers know their rights (presumably are far more able to afford lawyers) and do relatively well. Households with younger children and more children don't get more aid, as one would have suspected (and hoped).

The Odds: against

Education of mother

College 4 years or more	1.2-1
College 1-3 years	1.8-1
High school 4 years	2.4-1
High school 1-3 years	5.2-1
Elementary less than 9 years	8.0-1

The Odds: against

Race

White	2.2-1
Black	8.3-1
Spanish origin	5.1-1

No. of own children

1	3.3-1
2	2.6-1
3 or more	2.9-1

Age of oldest own child

Under 5 years	4.0-1
5-11 years	2.5-1
12-17 years	2.7-1

Source: Ibid.

WHAT ARE THE ODDS ON A SINGLE MOTHER RECEIVING CHILD SUPPORT BY PART OF THE COUNTRY?

One out of three West Coast mothers receive some child support, the "best" record in the country.

The Odds: against

Northeast	3.6-1
New England	3.4-1
North Central	2.6-1
East North Central	2.5-1
West North Central	3.0-1
South	3.3-1
South Atlantic	3.1-1
East South Central	3.3-1
West South Central	3.6-1
West	2.3-1
Mountain	2.5-1
Pacific	2.3-1

Source: Ibid.

IF YOU'RE DIVORCED AND BETWEEN THE AGES OF 26 AND 35 WHAT ARE THE ODDS ON REMARRYING BY NUMBER OF YEARS DIVORCED?

Three out of 10 who are divorced remarry within two years while about four out of 10 never remarry.

The Odds: Less than 1 year   5.1-1 for
            Less than 2 years   2.2-1 for
            Less than 3 years   1.4-1 for
            Less than 4 years   Even
            Less than 5 years   1.1-1 against
            Less than 10 years   1.5-1 against

Source: Department of Commerce, Bureau of the Census, *Number, Timing and Duration of Marriages and Divorces in the United States*, Series P-20, No. 297, 1976.

IF YOU REMARRY WHAT ARE THE ODDS BY AGE AND SEX ON YOUR SECOND MARRIAGE SUCCEEDING?

These days the more often you marry the greater the odds of divorcing again; twice-married women are more prone to redivorce than twice-married men. It's equally interesting to note that older men and women who are on second marriages have better odds of staying together the second time than even their one-time only married peers. Two reasons help explain this phenomenon. First, those that divorced in earlier years only to remarry faced stronger social pressures at the time. So presumably to remarry meant to work exceptionally hard to make the marriage work. Secondly, as you get older there's far less of an inclination to think that the grass is going to be greener in another pasture, having tried two of them. This is notably true of older women.

The Odds: against

Year of Birth	Men	Women
1900-1904	16.7-1	20.0-1
1905-1909	12.5-1	11.1-1
1910-1914	11.1-1	8.3-1
1915-1919	7.7-1	6.2-1
1920-1924	5.5-1	6.7-1
1925-1929	5.0-1	4.3-1
1930-1934	4.2-1	3.8-1
1935-1939	3.6-1	3.2-1
1940-1944	3.1-1	2.5-1
1945-1949	2.8-1	2.2-1

Source: Ibid.

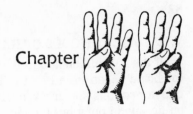

BEGINNINGS AND ENDINGS
Pregnancy, Birth and Abortion

In the time it will take to read this paragraph, three babies were born, one every 10 seconds. In the same amount of time, one pregnancy was aborted, or one every 30 seconds. Thus on a national basis, for every 1,000 live births there are 328 abortions.

Legalizing abortion has obviously had a big impact, though not so big an impact on the very poor who can't get government funding. Legalization has also made abortion dramatically safer. Of the almost four million abortions recorded in 1977 just 116 resulted in the death of the mother.

Over all our birth rate is declining. This is a direct reflection of the wide availability of birth control measures which has resulted in a marked trend toward deferring children. Proof? Twenty years ago the number of married women aged 20 to 24 who had never had a child was less than one in four (24%); in 1977 it was over four in 10 (43%) and still rising.

If you do wait to have a child the odds are with you all the way. Eight out of 10 women succeed in becoming pregnant within a year and the vast majority have uneventful, normal deliveries. But for the three out of four women who are in labor for over four hours the term "uneventful" will sound like a gross inaccuracy.

If you've waited till relatively late in life to get married and you're yearning to catch up on raising a big family, there's more good news. The older you are the shorter the odds on having twins. *Beginning and Endings* attempts to give you all the pregnant odds.

WHAT ARE THE ODDS OF HAVING A BUN IN THE OVEN?

Of the 255 million hundred weight of flour used annually 22% is used for home baked bread and rolls, with rolls accounting for about three quarters of this amount. We tried to get figures on pre-mixed dough and on pre-made buns but the industry keeps its bun facts close to its vest. This means the true odds are shorter than those shown.

The Odds: **5.1-1** against

Source: Data from Pillsbury, Inc.

WHAT ARE THE ODDS ON BEING A BAD EGG?

In 1978 5,596,000,000 dozen eggs were laid. During grading 167,880,000 dozen were rejected for breakage and other reasons.

The Odds: **32.3-1** against

Source: Data from the Department of Agriculture.

WHAT ARE THE ODDS ON NOT BEING BORN YESTERDAY?

In mid January, 1980 8,640 children were born per day and the total population was 222,059,633.

The Odds: **25,700-1** against

Source: Department of Commerce, Bureau of the Census.

WHAT ARE THE ODDS THAT THERE'S A SUCKER BORN EVERY MINUTE?

Excellent, in the spring of the year when thousands of suckers spawn their young in streams throughout North America every minute. During the balance of the year only a handful of suckers spawn under artificial conditions.

The Odds: In Spring   **1,000-1** for
            Balance of year   **100,000-1** against

Source: Data from U.S. Department of Fisheries.

WHAT ARE THE ODDS THAT YOU'LL CONCEIVE WITHIN ONE YEAR?

If you and your partner are both in good health, don't practice birth control, have regular sex and the woman is still in her 20's, head to the doctor if pregnancy does not occur within one year. The reason: 80 out of every 100 women will conceive within this period.

The Odds: **12.5-1** for

Source: Data from Northwestern University Medical School.

WHAT ARE THE ODDS ON HOW LONG A WOMAN WILL CARRY THE BABY BEFORE LIVE BIRTH?

The National Natality Survey gave the following breakdown in 1972.

The Odds: **Will have had baby by**

36 weeks	**10.0-1** against
39 weeks	**2.0-1** against
40 weeks	**14.0-1** for

Source: National Center for Health Statistics, *1972 National Natality Survey*, (presented at Population Association of America session: "How Safe are Childbearing and Birth Control?"), Atlanta, Georgia: April 13-15, 1978.

WHAT ARE THE ODDS ON HAVING A BABY BY TYPE OF DELIVERY?

Half of all mothers give birth spontaneously.

The Odds: against unless noted

Spontaneous	Even
Forceps	3.0-1
Caesarian	14.0-1
Breech	43.0-1
All other	104.0-1

Source: Ibid.

WHAT ARE THE ODDS THAT YOUR BABY WILL BE NORMAL?

In 1972 the National Center for Health Statistics did a survey of one of every 500 live births.

The Odds: **21.0-1** for

Source: Ibid.

WHAT ARE THE ODDS ON YOUR BABY BEING NORMAL BASED ON TYPE OF DELIVERY?

These odds vary dramatically. They're excellent for spontaneous and forceps births, disturbing for those delivered by Caesarian or as Breech babies. However these forms of delivery are most often the result of problems, not the cause.

The Odds: for

Spontaneous	209-1
Forceps	203-1
Caesarian	12-1
Breech	8-1
All other	27-1

Source: Ibid.

WHAT ARE THE ODDS ON DURATION OF LABOR?

Over three out of four prospective mothers endure more than three hours of labor pains.

The Odds:	0-3 hours	**3.3-1** against
	4-7 hours	**1.6-1** for
	8-11 hours	**12.1-1** for

Source: Ibid.

WHAT ARE THE ODDS ON COMPLICATIONS DURING LABOR?

The same study found that 20.2% of women had some form of complication during labor and delivery.

The Odds: **4.0-1** against

Source: Ibid.

WHAT ARE THE ODDS ON GIVING BIRTH WITHOUT AN ANAESTHETIC?

A stoic 203,000 women out of the 2,818,000 projected had no anaesthetic whatsoever.

The Odds: **13.3-1** against

Source: Ibid.

WHAT ARE THE ODDS THAT YOU'LL HAVE A MULTIPLE BIRTH?

The National Center for Health Statistics has produced a report on characteristics of births from 1973 to 1975 that holds some surprises on plural births. For example, the older the woman is the more apt she is to have a multiple birth. For every 1,000 white women giving live birth between the ages of 15 and 19 there are 12.1 multiple births. The figure more than doubles for whites between the ages of 35 and 39, to 26.2 multiple births per 1,000 live deliveries. At all ages blacks are about one third more apt to have multiple births than whites. This can be traced to parts of Africa where one in 20 to 30 deliveries is twins. Conversely, in Japan and China the rate is only one out of every 300.

Heredity accounts for more than two thirds of all twins. There are two kinds of twins: fraternal and identical. Fraternal twins grow from the fertilization of two separate eggs and are much more common in older women up to the age of 40. Identical twins occur when one fertilized egg splits into two equal parts. Unlike fraternal twins they are always the same sex.

The Odds:	On twins	**96-1** against	
	On triplets	**9,977-1** against	
	On quadruplets	**663,470-1** against	

Sources: Department of Health, Education, and Welfare, Public Health Service, National Center for Health Statistics, *Characteristics of Births, United States—1973-1975*, 1978, p. 16.

Department of Health, Education, and Welfare, National Center for Health Statistics, Public Health Service, *Multiple Births—United States—1964*, 1967, p. 3.

WHAT ARE THE ODDS THAT YOUR BABY WILL BE A CERTAIN BIRTH WEIGHT?

According to the National Natality Survey about two out of three babies weigh seven pounds 11 ounces or less at birth.

The Odds: against

5lb. 8 oz. or less	13.3-1	
5lb. 9oz.-6lb. 9oz.	4.8-1	
6lb. 10oz.-7lb. 11oz.	1.5-1	
7lb. 12oz.-8lb. 13oz.	2.8-1	
8lb. 14oz. or more	9.0-1	

The Odds: for

7lb. 11oz. or less	1.8-1 for

Source: National Center for Health Statistics, 1972 National Natality Survey, (presented at Population Association of America session: "How Safe Are Childbearing and Birth Control?", Atlanta, Georgia: April 13-15, 1978).

WHAT ARE THE ODDS THAT YOUR BABY WILL BE DELIVERED OUTSIDE A HOSPITAL?

Most women make it to the hospital. According to the 1977 Natality Statistics, of 3,326,632 births only 49,096 took place outside the hospital. For these a doctor was in attendance in one out of four cases, a midwife at another 25%. The other 50% is unspecified in the report but presumably involves some fairly ashen faced fathers, as well as cab drivers who wished they had passed up the fare.

The Odds: **65.6-1** against

Source: Department of Health, Education, and Welfare, Public Health Services, "Final Natality Statistics 1977", *Monthly Vital Statistics Report,* (PHS) No. 79-1120, Vol. 27, No. 11, Supplement, Hyattsville, Maryland: National Center for Health Statistics, 1979.

WHAT ARE THE ODDS THAT GIVEN A NUMBER OF CHILDREN A SPECIFIED NUMBER WILL BE BOYS AND/OR GIRLS?

If you're hoping for a boy or a girl the next time round this chart should make compelling reading. Clearly being able to get up an all girl or all boy family basketball team is some feat.

The Odds: against unless noted

1 boy	Even
1 girl	Even
2 boys or 2 girls	3.0-1
1 boy and 1 girl	Even
3 girls or 3 boys	7.0-1
2 boys and 1 girl or 2 girls and 1 boy	1.7-1
4 girls or 4 boys	15.1-1
3 girls and 1 boy or 3 boys and 1 girl	3.0-1
2 boys and 2 girls	1.7-1
5 boys or 5 girls	31.2-1
4 boys and 1 girl or 4 girls and 1 boy	5.4-1
3 girls and 2 boys or 3 boys and 2 girls	2.2-1

Source: James J. Nagle, *Heredity and Human Affairs,* St. Louis, C.V. Mosby Co., 1974, p. 228.

WHAT ARE THE ODDS ON HAVING AN ABORTION AS OPPOSED TO GIVING BIRTH BASED ON THE AGE OF THE MOTHER?

Among young teenagers there are 1,208 abortions for every 1,000 live births compared to an average of 328 per 1,000 live births among all age groups.

The Odds: against unless noted

15 and under	1.2-1 for
15-19	1.7-1
20-24	3.1-1
25-29	5.0-1
30-34	4.1-1
35-39	2.4-1
40	1.4-1

Source: Department of Health, Education, and Welfare, Center for Disease Control, *Abortion Surveillance 1976*, 1978.

WHAT ARE THE ODDS ON THE MARITAL STATUS OF SOMEONE HAVING AN ABORTION?

Of the 679,468 recorded abortions in 1976 where the marital status was known, 512,364 were administered to single women.

The Odds:	Married	**3.0-1** against
	Single	**3.0-1** for

Source: Ibid.

WHAT ARE THE ODDS ON DYING DURING AN ABORTION, OVER ALL AND BY WEEK OF GESTATION?

An analysis of 3,809,187 abortions shows 116 deaths. The risk climbs drastically week by week.

The Odds: against

Over all	33,331-1
8 weeks or less	166,667-1
9-10 weeks	58,523-1
11-12 weeks	35,713-1
13-15 weeks	12,820-1
16-20 weeks	6,210-1
21 + weeks	3,730-1

Source: Ibid.

WHAT ARE THE ODDS THAT A WOMAN WILL BECOME PREGNANT IF SHE OR HER PARTNER HAVE BEEN STERILIZED?

For every 10,000 men who have vasectomies 15 will subsequently father a child, while for certain types of female sterilization the failure rate is up to 2% of all operations.

The Odds: against

Vasectomy	666.0-1
Conventional	
Laparotomy	49.0-1
Laparascopy	49.0-1
Minilaparotomy	165.5-1
Colpotomy	1,817.2-1
Culdoscopy	1,817.2-1

Source: Population Information Program, *Population Reports*, "M/F Sterilization," Special Topic Monograph No. 2, Baltimore, Md., Johns Hopkins University, March 1978.

WHAT ARE THE ODDS THAT A WOMAN WILL BECOME INADVERTANTLY PREGNANT BASED ON THE CONTRACEPTIVE METHOD USED?

Over all, 11% of married women have an unwanted pregnancy while practicing some form of contraception during the first year of use. Clearly the rhythm method is least effective.

The Odds: against

Rhythm	51.6-1
Foam/cream/jelly	70.1-1
Diaphragm	91.1-1
Condom	109.4-1
IUD	280.4-1
Pill	602.7-1

Source: Barbara Vaughan, James Trussel, Jane Menken and Elise F. Jones, "Contraceptive Failure Among Married Women in the United States, 1970-1973", *Family Planning Perspectives*, Vol. 9, No. 6, 1977, pp. 251-258.

WHAT ARE A WOMAN'S ODDS ON HAVING AN ILLEGITIMATE BABY?

In the 35 years from 1940 to 1975 births out of wedlock increased more than fourfold, from 90,000 to 448,000 per year. This amounts to 14.3% of all annual births. By way of contrast over the same period the number of single women aged 14 to 44 only increased 39%, from 12.4 million in 1940 to 17.2 million in 1975.

The Odds:

Among all women giving birth	**6.0-1** against
Among all single women aged 14-44	**37.4-1** against

Source: Department of Commerce, Current Population Reports, *Perspectives on American Fertility*, Special Studies Series P-23, No. 70, 1978, pp. 39-43.

WHAT ARE THE ODDS, ACCORDING TO STATE, OF HAVING AN ABORTION?

These odds compare all females in the normal childbearing years of 15 to 44 with health agency reports of abortions on a state by state basis. In selected instances the figures reflect far more than a straightforward relationship of resident females in the fertile years to abortion. Thus, Washington, D.C.'s exceptional ratio doesn't take into account the many residents of Maryland and Virginia who had an abortion performed in the District (one reason Virginia's odds are so long). By the same token, New York has historically had far more liberal laws in this matter than neighboring states and the patient is anonymous in New York City. The result: inflated figures.

The Odds: against

Alabama	111.0-1	Maine	116.0-1
Alaska	74.2-1	Maryland	46.1-1
Arizona	94.8-1	Massachusetts	39.0-1
Arkansas	135.0-1	Michigan	48.8-1
California	33.7-1	Minnesota	63.9-1
Colorado	52.5-1	Mississippi	334.0-1
Connecticut	51.1-1	Missouri	80.8-1
Delaware	52.9-1	Montana	91.7-1
Dist. of Col.	4.4-1	Nebraska	84.8-1
Florida	47.0-1	Nevada	57.1-1
Georgia	43.9-1	New Hampshire	94.3-1
Hawaii	38.8-1	New Jersey	53.1-1
Idaho	189.0-1	New Mexico	51.9-1
Illinois	37.1-1	New York	26.1-1
Indiana	140.0-1	North Carolina	51.6-1
Iowa	115.0-1	North Dakota	77.4-1
Kansas	54.6-1	Ohio	64.5-1
Kentucky	87.0-1	Oklahoma	78.8-1
Louisiana	136.0-1	Oregon	41.0-1

The Odds: against

Pennsylvania	48.1-1	Vermont	46.2-1
Rhode Island	50.5-1	Virginia	495.0-1
South Carolina	112.0-1	Washington	35.3-1
South Dakota	91.9-1	West Virginia	397.0-1
Tennessee	55.2-1	Wisconsin	72.4-1
Texas	91.8-1	Wyoming	234.0-1
Utah	110.0-1		

Source: Department of Health, Education, and Welfare, Center for Disease Control, *Abortion Surveillance 1976*, (CDC) 78-8205.

Chapter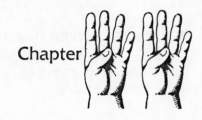

FROM BIRTH TO BRILLIANCE
Traits, Talents and Prospects

All of us think our children are special and so they are. In the pages which follow you'll be able to determine just how special your children are and in which ways.

We start the way mothers start, by scrutinizing baby's behavior. If he's walking by 10 months he's clever, or at least coordinated, while if he waits to 15 months 10 contemporaries beat him to his feet.

Next we examine schooling. This shows private and parochial schools are on the decline. It also shows that in the world's best educated society there's still a lot of wastage along the way. For every 100 children entering grade school only 88 graduate, while another 13 drop out of high school. So one out of four kids starts life with a crippling lack of education.

We consider the odds on going to college. They're good: very nearly even. What's more, among those going most get into the college of their first choice. This reflects realism in their selection of higher institutions because fewer than 4% of those tested on the SAT's score 650 or over.

Because the SAT's are such a looming hurdle for all college-bound teenagers we figured the odds on achieving certain scores and traits that make up the typical entering freshman.

We delved into distance from home (most are close), financial assistance (most need it), where students want to live (men are bigger on co-ed dormitories than women) and the need for remedial study.

While there are good colleges and universities all over the country, educators rate a few as particularly fine, so we profiled the odds on admissions to these separately.

Finally we looked at individual courses of study, emphasizing the practical, as well as the odds of going to graduate school and which ones. At about three admissions per 100 applicants Harvard Medical School is still a tough nut to crack.

WHAT ARE THE ODDS OF BEING THE REAL McCOY?

Of the 239,927,977 Social Security Administration records on individuals there are 102,894 people named McCoy, so the odds are in terms of being the real McCoy.

The Odds: **232,839-1** against

Source: Department of Health, Education, and Welfare, Social Security Administration, *Report of Distribution of Surnames in the Social Security Number File,* (7-75) (BDP) No. 034-75 1974.

WHAT ARE THE ODDS ON BEING A GIRL FRIDAY?

We compared girls born on other days of the week to those born on Friday.

The Odds: **6.0-1** against

Source: Department of Commerce, Bureau of the Census.

WHAT ARE THE ODDS ON KEEPING UP WITH THE JONESES?

Of all Social Security Administration names 1,331,205 are for people named Jones. Therefore, the remaining 238,596;772 may be said to be keeping up with the Joneses.

The Odds: **179.2-1** for

Source: Department of Health, Education, and Welfare, Social Security Administration, *Report of Distribution of Surnames in the Social Security Number File,* (BDP) No. 034-75 (7-75) 1974..

YOUR CHILD'S TRAITS AND TALENTS

Virtually all of us think our new children are special. The Denver Development Screening Test is a series of tests designed to evaluate young children by means of standardized screening procedures. Most experts agree that these tests are one of the better standardized measurements available. The one weakness is that it measures kids against one another in just one town, Denver, Colorado, which may not be typical of the entire United States.

The Odds: **That your baby will laugh**

By 6 weeks	3-1 against
By 8 weeks	Even
By 10 weeks	9-1 for

The Odds: **That your baby will sit up without support**
By 19 weeks	3-1 against
By 22 weeks	Even
By 26 weeks	3-1 for
By 31 weeks	9-1 for

The Odds: **That your baby will walk without assistance**
By 11 months	3-1 against
By 12 months	Even
By 13½ months	3-1 for
By 14½ months	9-1 for

The Odds: **That your baby will say "dada" or "mama"**
By 5½ months	3-1 against
By 7 months	Even
By 9 months	3-1 for
By 10 months	9-1 for

The Odds: **That your child will put together his first fully formed sentence**
By 14 months	3-1 against
By 20 months	Even
By 22 months	3-1 for
By 2½ years	9-1 for

The Odds: **That your child can give his first and last names**
By 2 years	3-1 against
By 2¾ years	Even
By 3¼ years	3-1 for
By 3¾ years	9-1 for

The Odds: **That your child can draw a person in three recognizable parts**
By 3¼ years	3-1 against
By 4 years	Even
By 4½ years	3-1 for
By 5½ years	9-1 for

The Odds: **That your child has learned to peddle a tricycle**
By 21 months	3-1 against
By 2 years	Even
By 2¾ years	3-1 for
By 3 years	9-1 for

Source: Denver Development Screening Test, LADOCA Foundation, Denver, Colorado.

WHAT ARE THE ODDS OF HAVING AN AUTISTIC CHILD?

One out of every 3,000 children has some form of autistic disorder.

The Odds: **2,999-1** against

Source: National Institute of Mental Health, *Summary of the Statistical Findings of the President's Commission on Mental Health,* 1977.

WHAT ARE THE ODDS THAT YOUR CHILD WILL HAVE TO WEAR CORRECTIVE LENSES?

In 1971, the National Center for Health Statistics reported that of their sample of 191,602 children between the ages of three and 16, the number who had to wear corrective lenses was 94,284. This amounted to 44.2% of the boys and 53.8% of the girls.

The Odds: Boys **1.3-1** against
Girls **1.2-1** for

Source: National Center for Health Statistics, Vital and Health Statistics, Series 10, No. 79; and unpublished data derived from the Department of Commerce, Bureau of the Census, *Statistical Abstract,* 1978, p. 118.

WHAT ARE THE ODDS ON PARENTS USING DIFFERENT PHYSICAL FORMS OF VIOLENCE WHEN DEALING WITH THEIR CHILD?

America may have lost most of its woodsheds but spanking has by no means disappeared. Fifty-eight percent of parents still use this disciplinary measure. Nothing startling there. What is startling is that just over one in 20 children has been kicked, bitten, hit with a fist or just plain beaten. One child in 500 has either been threatened by a knife or gun or actually cut or shot.

The Odds: against unless noted

Slapped or spanked	1.4-1 for
Pushed, grabbed or shoved	1.5-1
Hit with something	6.5-1
Threw something	17.5-1
Kicked, bit or hit with fist	30.2-1
Beat up	75.9-1
Threatened with a knife or gun	999.0-1
Used a knife or gun	999.0-1

WHAT ARE THE ODDS THAT CHILD ABUSE WILL GO UNREMEDIED?

A 1978 analysis of actions taken after reports of child abuse and neglect in 33 different United States locales showed that in 14.% of all cases nothing is done.

The Odds: **5.8-1** against

Source: American Humane Association, *National Analysis of Official Child Neglect and Abuse Reporting.*

WHAT ARE THE ODDS OF YOUR CHILD BEING TAKEN INTO POLICE CUSTODY?

In 1976 the police picked up 1,569,626 juvenile offenders out of a possible 56,245,000 families with children during that year. In just over half of these arrests (53.4%) the case was referred to a juvenile court while 39% of the total were simply released.

It's impossible to determine how often the same juvenile was arrested or how many came from the same family, so the odds which follow are shorter than reality (probably by as much as 50%).

The Odds: **34.8-1** against

Source: Department of Justice, *F.B.I. Uniform Crime Report,* 1976.

WHAT ARE THE ODDS ON YOUR CHILD ATTENDING A PRIVATE ELEMENTARY OR SECONDARY SCHOOL?

Of the 49,139,000 students enrolled in 1976-77, 4,804,000 were in private institutions.

The Odds: **9.3-1** against

Source: Department of Health, Education and Welfare, National Center For Education, *Digest of Education Statistics 1977-78,* 1978.

IF YOU'RE A CATHOLIC WITH SCHOOL AGE CHILDREN WHAT ARE THE ODDS THAT THEY'LL GO TO A PAROCHIAL SCHOOL?

In 1977 Catholic families had 14,241,000 children between the ages of five and 17. Of these, 3,365,000 were in parochial schools.

The Odds: **3.2-1** against

Source: *Standard Education Almanac 1978-1979,* Chicago, Illinois, Marquis Academic Media, pp. 201, 202.

WHAT ARE THE ODDS OF YOUR CHILD GRADUATING FROM GRAMMAR SCHOOL?

Out of every 100 children in the fifth grade, 98 will enter the ninth grade.

The Odds: **7.3-1** for

Source: U.S. Department of Health, Education and Welfare, National Center for Education Statistics, *Digest of Education Statistics*, 1979, p. 14.

WHAT ARE THE ODDS THAT YOUR CHILD WILL GRADUATE FROM HIGH SCHOOL?

Of all kids enrolled in the fifth grade 75% graduate from high school, while of those that enter high school 76% finish.

The Odds:

Among all 5th graders	**3.0-1** for	
Among those who enter high school	**3.1-1** for	
Among those who enter 11th grade	**10.0-1** for	

Source: Ibid.

WHAT ARE THE ODDS THAT YOUR CHILD WILL GO TO COLLEGE?

Currently about 47% of all young adults start some form of study at a college or university.

The Odds: **1.1-1** against

Source: Ibid.

WHAT ARE THE ODDS THAT A COLLEGE FRESHMAN'S PARENTS WENT TO OR COMPLETED COLLEGE?

Half of all the fathers and about 40% of the mothers attended college.

The Odds:

	Male		Female	
Father	**Even**		**Even**	
Mother	**1.6-1**	against	**1.5-1**	against

Source: *The American Freshman: National Norms For Fall 1978*, Cooperative Institutional Research Program, American Council on Education, Los Angeles, University of California, pp. 23, 24, 39, 40.

WHAT ARE THE ODDS ON COLLEGE FRESHMAN MEN AND WOMEN HAVING CERTAIN CHARACTERISTICS?

Of all those entering college about nine out of 10 are white, almost two out of three men and over three out of four women had a B average or better, and half of all students have parents making between $10,000 and $25,000 each year.

The Odds: against unless noted

Race	Male	Female
White	8.3-1 for	7.1-1 for
Black	12.9-1	10.1-1
American Indian	124.0-1	124.0-1
Oriental	82.3-1	89.9-1
Mexican-American/Chicano	99.0-1	99.0-1
Puerto Rican-American	110.0-1	124.0-1
Other	51.6-1	65.7-1

Average Grade In High School	Male	Female
A or A +	10.8-1	7.1-1
A –	8.5-1	5.7-1
B +	4.7-1	3.4-1
B	2.9-1	2.7-1
B –	5.8-1	8.3-1
C +	6.5-1	12.0-1
C	10.1-1	20.7-1
D	199.0-1	499.0-1

Parental Income	Male	Female
Less than $3,000	37.5-1	29.3-1
$3,000-$3,999	65.7-1	42.5-1
$4,000-$5,999	33.5-1	26.8-1
$6,000-$7,999	30.3-1	21.7-1
$8,000-$9,999	23.4-1	19.4-1
$10,000-$12,499	11.8-1	10.1-1
$12,500-$14,999	9.9-1	10.0-1
$15,000-$19,999	4.9-1	5.4-1
$20,000-$24,999	4.8-1	5.5-1
$25,000-$29,999	8.8-1	9.6-1
$30,000-$34,999	11.5-1	12.3-1
$35,000-$39,999	20.7-1	20.7-1
$40,000-$49,999	21.2-1	22.8-1
$50,000 +	12.7-1	13.9-1

Source: *The American Freshman: National Norms for Fall 1978*, Cooperative Institutional Research Program, American Council on Education, Los Angeles, University of California, pp. 31, 32, 16, 17.

WHAT ARE THE ODDS OF GETTING CERTAIN SAT SCORES ON THE VERBAL SECTION?

In 1979 close to one million students took the Scholastic Aptitude Tests (SAT's). Of this group only 1,416 males and 1,234 females scored between 750 and 800 on the verbal test, while just 3.7% of males and 3.1% of females scored higher than 650.

The Odds: against

Score	Male	Female	Total
750-800	337.0-1	414.2-1	373.2-1
700-749	86.9-1	106.2-1	95.9-1
650-699	41.8-1	50.8-1	46.1-1
600-649	21.2-1	23.7-1	22.4-1
550-599	16.6-1	14.0-1	13.3-1
500-549	7.4-1	7.9-1	7.6-1
450-499	5.6-1	5.8-1	5.7-1
400-449	4.5-1	4.4-1	4.5-1
350-399	5.1-1	4.9-1	5.0-1
300-349	7.2-1	6.7-1	6.9-1
250-299	13.0-1	11.9-1	12.4-1
200-249	25.0-1	22.4-1	23.6-1

Source: Educational Testing Service, Princeton, New Jersey, 1979.

WHAT ARE THE ODDS OF ACHIEVING PARTICULAR SAT VERBAL SUBSCORES?

As the figures demonstrate it's far harder to get top scores in reading comprehension than in the vocabulary section of the test.

The Odds: against

Score	Reading comprehension	Vocabulary
75-80	453.2-1	162.3-1
70-74	133.3-1	69.0-1
65-69	37.2-1	41.0-1
60-64	21.5-1	21.8-1
50-59	4.5-1	4.0-1
40-49	2.1-1	2.2-1
30-39	2.3-1	2.9-1
20-29	8.1-1	5.8-1

Source: Educational Testing Service, Princeton, New Jersey, 1979.

WHAT ARE THE ODDS OF GETTING CERTAIN SCORES ON THE SAT MATH SECTION?

When it comes to math the men outdistance the women by miles. More than 53,000 men scored over 650 in 1979. That's nearly three times the women's total of just under 19,000.

The Odds: against

Score	Male	Female	Total
750-800	60.6-1	348.3-1	108.4-1
700-749	30.0-1	102.6-1	45.1-1
650-699	15.5-1	39.5-1	22.7-1
600-649	9.0-1	16.8-1	11.9-1
550-599	7.0-1	10.1-1	8.3-1
500-549	5.3-1	6.1-1	5.7-1
450-499	6.0-1	5.5-1	5.7-1
400-440	6.8-1	5.2-1	5.8-1
350-399	8.4-1	5.7-1	6.8-1
300-349	11.5-1	6.8-1	8.5-1
250-299	24.0-1	13.0-1	16.8-1
200-249	135.4-1	74.3-1	95.1-1

Source: Educational Testing Service, Princeton, New Jersey, 1979.

WHAT ARE THE ODDS ON GOING TO A COLLEGE OR UNIVERSITY BASED ON THE NUMBER OF MILES FROM HOME?

About half of all students go to an institution which is within 50 miles of their home. Only one male in 11 and one female in 13 go more than 500 miles away.

The Odds: against

Miles From Home	Male	Female
5 or less	8.8-1	9.2-1
6-10	7.1-1	7.6-1
11-50	2.8-1	2.7-1
51-100	6.0-1	5.4-1
101-500	2.5-1	2.6-1
500 +	10.2-1	11.9-1

Source: *American Freshman: National Norms For Fall 1978,* Cooperative Institutional Research Program, American Council on Education, Los Angeles, University of California, pp. 29, 45.

WHAT ARE THE ODDS ON GETTING INTO THE COLLEGE OR UNIVERSITY OF YOUR FIRST CHOICE?

Over three out of four college bound seniors are accepted by the school of their choice.

The Odds: against unless noted

	Male	Female
First choice	3.0-1 for	3.3-1 for
Second choice	4.4-1	4.4-1
Third choice	21.7-1	26.8-1
Less than third choice	49.0-1	70.4-1

Source: *The American Freshman: National Norms for Fall 1978,* Cooperative Institutional Research Program, American Council on Education, Los Angeles, University of California, pp. 18, 34.

WHAT ARE THE ODDS ON SEEKING FINANCIAL ASSISTANCE WHEN APPLYING TO COLLEGE?

Fully four out of five college bound students who take the SAT's intend to seek financial assistance. Here's how it breaks down by ethnic group.

The Odds: for

Black	17.8-1
Mexican-American	11.5-1
Puerto Rican	10.1-1
Oriental	8.7-1
American Indian	7.2-1
Other	6.6-1
White	4.1-1

Source: Educational Testing Service, Princeton, New Jersey, 1979.

WHAT ARE THE ODDS ON PLANNED HOUSING PREFERENCE WHEN AT COLLEGE?

The Odds: against

	Male	Female
At home	3.2-1	2.7-1
Single sex dorm	4.8-1	2.5-1
Coed dorm	1.6-1	2.6-1
Fraternity & sorority	22.8-1	25.3-1
On-campus apartment	9.4-1	14.1-1
Off-campus apartment	12.3-1	16.8-1

Source: Educational Testing Service, Princeton, New Jersey, 1979.

WHAT ARE THE ODDS, BY TOPIC, ON INCOMING MALE AND FEMALE FRESHMEN NEEDING REMEDIAL WORK?

In virtually all subjects girls are better prepared than boys when they enter college.

The Odds: against

	Male	Female
English	5.3-1	7.3-1
Reading	10.0-1	13.1-1
Mathematics	3.6-1	2.6-1
Social sciences	27.6-1	21.2-1
Science	8.9-1	5.4-1
Foreign language	5.4-1	6.8-1

Source: *The American Freshman: National Norms For Fall 1978*, Cooperative Institutional Research Program, American Council on Education, Los Angeles, University of California, pp. 16, 33.

WHAT ARE THE ODDS ON COMPLETING A FOUR YEAR COLLEGE COURSE?

Of all those who entered college in the fall of 1977, 51% will earn a bachelor's degree in 1981.

The Odds: Just over **even**

Source: College Career Placement offices

AMONG MALE CANDIDATES FOR A BACHELOR'S DEGREE, WHAT ARE THE ODDS ON THE FIELD THEY WILL CHOOSE?

Over one out of five students majors in business. This is little changed from five years ago. What is changed is interest in majoring in the social sciences. This went down from 20.6% of all 1970-71 degrees to 15.6% of all 1975-76 degrees.

The Odds: against

Field	
Business and management	3.4-1
Social sciences	5.4-1
Biological physical sciences	8.5-1
Engineering	10.2-1
Education	11.0-1
Psychology	21.2-1
All other fields	3.4-1

Source: National Center for Education Statistics, *Digest of Education Statistics*, 1979, Washington, D.C., p. 117.

WHAT ARE THE ODDS OF GOING ON TO GRADUATE SCHOOL FROM 12 TOP COLLEGES?

An astounding 86% of Harvard/Radcliffe graduates continue their education. Even at the lowest scoring school in this category, Stanford, one out of two students goes to grad school.

The Odds: for unless noted

Harvard/Radcliffe	6.1-1
Rice	2.3-1
Williams	2.3-1
Wesleyan	2.3-1
Duke	2.1-1
M.I.T.	2.0-1
University of Chicago	1.8-1
Yale	1.6-1
Princeton	1.5-1
Cal Tech	1.3-1
Bryn Mawr	1.2-1
Stanford	Even

Source: College Career Placement offices

WHAT ARE THE ODDS OF BEING ACCEPTED AS AN UNDERGRADUATE BY CERTAIN TOP COLLEGES?

We chose this group of top colleges on the basis of two factors: over 60% of the freshman class achieved 600+ SAT scores on both their verbal and math exams, and university educators consistently rank these schools as the best in the United States.

Colleges traditionally accept from one and a half to two students for every one that matriculates. These figures reflect acceptances compared to the total number of applicants for 1978.

The figures have to be treated with some caution. The reputations of the schools deter many students from applying, as do application fees. So if you have three would-be scientists at home it doesn't mean that only one out of the three will make Cal Tech.

Duke and Amherst are the toughest to get into, while applicants to Vassar, the University of Chicago, and Smith don't face such stiff competition. But the odds against males at Smith are astronomic; it's not co-ed.

The Odds: against unless noted

Amherst	5.5-1
California Institute of Technology	2.0-1
Columbia	2.1-1

The Odds: against unless noted

Cornell	1.7-1
Dartmouth	3.3-1
Harvard/Radcliffe	4.6-1
M.I.T.	1.3-1
Princeton	3.3-1
Rice	3.0-1
Smith	1.2-1 for
Stanford	3.0-1
Swarthmore	1.4-1
University of Chicago	1.5-1 for
Wellesley	Even
Wesleyan	2.3-1
Williams	3.5-1
Vassar	1.2-1 for
Yale	3.0-1

Source: Admissions offices

WHAT ARE THE ODDS OF DIFFERENT TYPES OF COLLEGE GRADUATES WORKING IN A FIELD DIRECTLY RELATED TO THEIR MAJOR?

Predictably the more specialized their training the more apt they are to land a job in a pertinent field. Surprisingly even in the humanities the odds are their job will relate to their major. Because they are old, these figures must be treated with caution. For example, education majors will have a much harder time as the student population decreases.

The Odds: for unless noted

Sex

Men	2.0-1
Women	2.5-1

Type of Degree

Bachelor's	1.6-1
All other	6.4-1

Major Field of Study

Business & commerce	1.6-1
Education	4.5-1
Humanities	1.3-1
Social science	1.2-1 against
All other	3.1-1

Source: Department of Labor, Bureau of Labor Statistics, *Employment of Recent College Graduates*, October, 1972.

AMONG FEMALE CANDIDATES FOR A BACHELOR'S DEGREE WHAT ARE THE ODDS ON THE FIELD THEY WILL CHOOSE?

Changing needs and opportunities in our society are clearly reflected in the figures for women. While education is still the largest field (26.8%), it's down about 10 percentage points from 1970-71 figures. At the same time business and management has grown rapidly, and health professions have almost doubled.

The Odds: against

Field	
Education	2.7-1
Social sciences	7.8-1
Health professions	8.9-1
Business and management	13.9-1
Psychology	14.6-1
Fine & applied arts	15.4-1
All other fields	2.1-1

Source: National Center for Education Statistics, *Digest of Education Statistics,* 1979, Washington, D.C., p. 117.

WHAT ARE THE ODDS ON BEING ACCEPTED AT SELECTED TOP GRADUATE SCHOOLS?

As the figures below show Harvard and Yale live up to their reputations as tough places to get into. It is surprising that Stanford attracts far more business school applicants for the number of places it offers than the Harvard Business School.

The Odds: against

Selected Top Medical Schools

Harvard	32.3-1
Yale	27.6-1
Stanford	24.0-1

Selected Top Business Schools

Stanford	7.7-1
University of Chicago	4.3-1
Northwestern	4.1-1
Harvard	4.0-1
University of Pennsylvania	3.4-1

Selected Top Law Schools

Yale	8.1-1
Harvard	7.3-1
Stanford	6.2-1
Michigan	3.8-1

Source: Admissions offices

Chapter

IN SICKNESS AND IN HEALTH
Our Bodies/Our Minds

Women's liberation apart, women do look up to men; at an average of five feet six inches, they're very nearly four inches shorter. And at an average of 160 pounds, they're about 30 pounds lighter than their male counterparts.

In Sickness and In Health examines the multi-faceted world of our bodies and our mental health. There are no end of surprises in the pages which follow. You'll discover that the TV image of the trim executive and his family is a myth. Wealth brings girth.

You'll see that you're three and a half times more apt to contract chicken pox than measles and that mumps have gone out of fashion. But where you live has much to do with what you'll catch.

You'll also find that disease is sexist. Men are far more apt to have skin problems than women are, while women are far more prone to everything from arthritis to mental disease.

The odds which follow demonstrate another home truth. Sickness, like sex, starts in the mind. The separated and divorced are physically ill far more often than others, while mental illness is most common among those who have never married or aren't married now.

The news isn't all bad. False teeth are on the way out. Your odds of coming out of the hospital alive are good (and one out of ten of us go in every year). Most surgery is remarkably safe these days. And we struggle and prevail rather than end it all; the odds on suicide are long whatever your age.

The hypochondriacs among us will, no matter what the number, identify with the "to one" side of the statistic. For the rest of us it's reassuring to know we're getting better all the time.

WHAT ARE THE ODDS THAT A MAN WILL BE A CERTAIN HEIGHT?

Half of all men are taller than 5'9" while about one in 10 tops six feet.

The Cumulative Odds:

Height range (inches)	
64.4 or less	19.0-1 against
64.5 to 65.5	9.0-1 against
65.6 to 67.1	3.0-1 against
67.2 to 69	Even
69.1 to 70.8	3.0-1 for
70.9 or over	3.0-1 against

Source: Department of Health, Education and Welfare, National Center for Health Statistics, *Advancedata*, No. 3, Rockville, Maryland, 1976.

WHAT ARE THE ODDS THAT A MAN WILL BE A CERTAIN WEIGHT?

The odds are even that a man will weigh 170 pounds or less.

The Cumulative Odds:

Weight (pounds)	
128 or less	19.0-1 against
129–137	9.0-1 against
138–152	3.0-1 against
153–170	Even
171–189	3.0-1 for
190 or over	3.0-1 against

Source: Ibid.

WHAT ARE THE ODDS THAT A WOMAN WILL BE A CERTAIN HEIGHT?

One woman in four is five feet tall or less. Half of all women are under five feet two while one in 10 is five feet six inches or taller.

The Cumulative Odds:

Height range (inches)	
59.5	19.0-1 against
60.5	9.0-1 against
62.0	3.0-1 against
63.7	Even
65.3	3.0-1 for
66.8 or over	3.0-1 against

Source: Ibid.

WHAT ARE THE ODDS THAT A WOMAN WILL BE A CERTAIN WEIGHT?

One in four women weighs 122 pounds or less while for those aged 18 to 24 (not shown) this figure is 114 pounds. Half of all women weigh 137 pounds or less.

The Cumulative Odds:

Weight (pounds)	
104 or less	19.0-1 against
105-110	9.0-1 against
111-122	3.0-1 against
123-137	Even
138-159	3.0-1 for
160 or over	3.0-1 against

Source: Ibid.

WHAT ARE THE ODDS BY SEX, AGE, INCOME AND RACE THAT YOU'LL BE OVERWEIGHT?

One out of two women over the age of 17 believe they're overweight. Whites tend to have greater perceived weight problems than other races. Weight problems increase as incomes increase; the same is true of age until you're 65.

The Odds: against unless noted

	Men			Women		
	All races	White	Black	All races	White	Black
17 & over	2.3-1	2.1-1	4.1-1	Even	Even	1.3-1
17-44	2.6-1	2.4-1	4.4-1	1.1-1	Even	1.2-1
45-64	1.6-1	1.5-1	3.4-1	1.3-1	1.3-1 for	1.1-1 for
65 +	3.2-1	3.1-1	4.8-1	1.7-1	1.6-1	3.2-1

Income

Less than			
$5,000	3.9-1		1.3-1
$5,000-$9.	2.7-1		Even
$10.-$14.9	2.1-1		1.1-1 for
$15,000 +	1.7-1		Even

Source: Ibid.

WHAT ARE THE ANNUAL ODDS BY REGION ON CATCHING VARIOUS CONTAGIOUS DISEASES?

In 1977 the Center for Disease Control reported 188,396 cases of chicken pox, 57,345 of measles and 21,436 of mumps. Chicken pox ruled the roost in New England, while the West North Central states were the spots for measles and mumps.

The Odds:　against

	Chicken Pox	Measles	Mumps
United States	1,147-1	3,771-1	10,091-1
New England	576-1	4,840-1	16,300-1
Middle Atlantic	1,678-1	4,276-1	24,318-1
East North Central	548-1	2,926-1	5,530-1
West North Central	731-1	1,698-1	3,239-1
South Atlantic	3,092-1	7,131-1	34,476-1
East South Central	4,326-1	6,566-1	11,042-1
West South Central	2,247-1	9,574-1	11,613-1
Mountain	1,341-1	4,179-1	14,558-1
Pacific	1,843-1	2,754-1	16,975-1

Source: Department of Health, Education, and Welfare, Center for Disease Control, MMWR, *Annual Summary 1977*, September 1978, Vol. 26, No. 53.

WHAT ARE THE ODDS THAT A YOUNG CHILD WILL BE IMMUNIZED AGAINST SPECIFIC DISEASES?

In 1977 among children under the age of four, between 30 and 50% were not immunized against any given contagious diseases.

The Odds:　for unless noted

Diptheria-Tetanus-Pertuis (3 shots +)	2.3-1
Polio (3 doses +)	1.5-1
Measles	1.7-1
Rubella	1.5-1
Mumps	1.1-1 against

Source: U.S. Center for Disease Control, *United States Immunization Survey*, Atlanta, Georgia, 1977.

WHAT ARE THE ODDS OF BEING IN STITCHES?

In 1978 a total of 1,320,500 people worked in the garment industry.

The Odds:　**164.0-1** against

Source: Data from Department of Labor, Bureau of Labor Statistics.

BASED ON SEX AND AGE, WHAT ARE THE ODDS ON HAVING ONE OR MORE SIGNIFICANT SKIN CONDITIONS?

We think of acne and other major skin conditions as a teenage phenomenon. These figures show that the problem may start in these years but it doesn't end when they're over. Throughout their post-teen lives over one out of three Americans has at least one significant skin condition.

The Odds: against

All men	1.9-1	25-34 years	2.1-1
All women	2.5-1	35-44 years	2.0-1
1-5 years	6.0-1	45-54 years	1.8-1
6-11 years	4.7-1	55-64 years	1.8-1
12-17 years	1.8-1	65-74 years	1.4-1
18-24 years	1.7-1		

Source: Department of Health, Education, and Welfare, *Advancedata* No. 4, Rockville, Maryland, 1977.

WHAT ARE THE ODDS ON A TEENAGER SUFFERING FROM ACNE?

A recent study shows that acne affects some 60 million persons in the United States including 95% of all teenagers.

The Odds: **19.0-1 for**

Source: *International Market Research Report No. 562,* Frost and Sullivan, Inc., New York.

WHAT ARE THE ODDS OF WEARING GLASSES AND, IF SO, OF BEING NEARSIGHTED OR FARSIGHTED?

Just over half the population between the ages of six and 74 (52.4%) wear glasses.

The Odds:

Of wearing glasses	**1.1-1** for
If so of being nearsighted	**1.1-1** against
If so of being farsighted	**1.2-1** against
If so other eye problem	**14.0-1** against

Source: Department of Health, Education and Welfare, National Center for Health Statistics, *Refraction Status and Motility Defects of Persons 4-74 Years, United States 1971-1972,* (PHS) 78-1654, Series II, No. 206, Hyattsville, Maryland, August 1978.

WHAT ARE THE ODDS OF SUFFERING FROM FLAT FEET?

The only figure found doesn't include a prime target group: the military. The odds which follow are based on the civilian, non-institutionalized population.

The Odds: **62.4-1** against

Source: *Journal of the American Podiatry Association*, Vol. 67, No. 2, February 1977.

WHAT ARE THE ODDS, ACCORDING TO AGE AND SEX, ON HAVING PERIODONTAL DISEASE?

As we grow older gum disease becomes a fact of life for most of us. This is far more true of men than of women. Between the ages of 18 and 44 half of the men have this mouth problem while only about one in three of the women are similarly affected.

The Odds: against unless noted

Age	Men	Women
6-11	5.0-1	8.1-1
12-17	1.6-1	2.6-1
18-44	Even	2.0-1
45-64	1.6-1 for	1.1-1 for
65-74	2.6-1 for	1.3-1 for

Source: National Center for Health Statistics, Department of Health, Education and Welfare, *The Natural History of Periodontal Disease in Man*, Unpublished, 1978.

WHAT ARE THE ODDS OF BEING HOSPITALIZED DURING A GIVEN YEAR?

Excellent if you're one of the roughly one and a half million babies born each year. Curiously, the government derives its statistics on the basis of discharges. So we looked at them in this optimistic way as well.

The Odds: **6.1-1** against

Source: Department of Health, Education, and Welfare, Utilization of Short-Stay Hospitals: Summary of the United States, 1976, Series 13, Number 37, No. (PHS) 78-1788, Hyattsville, Maryland, June 1978.

WHAT ARE THE ODDS THAT YOU'LL SUFFER FROM SOME FORM OF COMMON DISEASE THIS YEAR?

The odds are better than even that each of us will have some form of upper respiratory infection this year. Women suffer far more than men from these. For every 100 women there are 64.7 acute conditions. Over four out of 10 of these illnesses require medical attention. By the same token about twice as many women as men suffer from headaches each year.

The Odds: against

	Male	Female
Common cold	1.2-1	1.1-1
Influenza	1.7-1	1.3-1
Disease of ear	14.9-1	11.2-1
Genitourinary disorder	65.7-1	9.0-1
Fracture or dislocation	25.3-1	33.5-1
Dental condition	29.3-1	33.5-1
Migraine headache	57.8-1	28.4-1

Source: Department of Health, Education and Welfare, National Center for Health Statistics, *Acute Conditions*, No. (PHS) 78-1553, Hyattsville, Maryland.

WHAT ARE THE ODDS BY AGE ON A WOMAN HAVING AT LEAST ONE GYNECOLOGICAL PROBLEM OTHER THAN BREAST TROUBLE EACH YEAR?

These are the most prevalent in the years between 17 and 44 when 47 out of every thousand families suffer from at least one such complaint.

The Odds: against

All women	35.8-1
Under 17	249.0-1
17-44	20.4-1
45-64	30.8-1
65 +	75.3-1

Source: Department of Health, Education and Welfare, *Prevalence of Chronic Conditions of the Genitourinary, Nervous, Endocrine, Metabolic, and Blood-Forming Systems and of Other Selected Chronic Conditions, United States – 1973*, No. (HRA) 77-1536, Series 10, No. 109, Rockville, Maryland, 1977.

WHAT ARE THE ODDS BY AGE ON HAVING CERTAIN DISEASES?

At all ages deafness is the most prevalent form of disability shown. The disease is 25% more prevalent among women and almost twice as common among non-whites as among whites.

The Odds: against

	All ages	Under 17	17-44	45-64	65 +
Hearing impairment	13.0-1	75.9-1	22.6-1	7.8-1	3.3-1
Diabetes	48.0-1	768-1	111-1	22.5-1	11.7-1
Thyroid condition	20.9-1	908-1	61.1-1	37.0-1	49.8-1
Anemia	68.0-1	124-1	53.9-1	68.4-1	46.8-1
Ulcer-stomach	57.1-1	91.6-1	91.6-1	28.9-1	33.5-1

Source: Department of Health, Education and Welfare, *Prevalence of Selected Impairments, United States—1971*, No. (HRA) 75-1526, Series 10, No. 99, Rockville, Maryland, May 1975.

WHAT ARE THE ODDS BY TYPE ON HAVING AN OPERATION IN A GIVEN YEAR?

Out of every 100,000 people about 9,500 will have an operation in any given year, and one out of five of these will be gynecological.

The Odds: against

Any operation	9.5-1
Gynecological surgery (any form)	53.9-1
Abdominal surgery	74.0-1
Orthopedic surgery	78.7-1
Hysterectomy	160.0-1
Plastic surgery	194.0-1
Caesarean	287.0-1
Tonsillectomy	334.0-1
Dental surgery	552.0-1

Source: Department of Health, Education and Welfare, *Utilization of Short Stay Hospitals*, Annual Summary 1976, No. (PHS) 78-1788, June 1978.

BASED ON MARITAL STATUS WHAT ARE THE ODDS ON BEING SERIOUSLY ILL PER YEAR FOR WOMEN?

Separated and divorced women are far more apt to be seriously ill than other women. They also visit the doctor more often (6.8 times per year versus an all female average of 5.4) and have far more days when their activity is restricted due to illness (29.7 and 26.4 versus an 18.8 day average).

The Odds: for

All women 17 +	174-1
Married women	174-1
Widowed women	165-1
Separated women	223-1
Divorced women	216-1
Never married women	161-1

Source: Department of Health, Education and Welfare, *Differentials in Health Characteristics by Marital Status, United States, 1971-1972*, No. (HRA) 76-1531, Series 10, No. 104, Rockville, Maryland, March 1976.

WHAT ARE THE ODDS ON HAVING A HEART ATTACK?

The American Heart Association estimates that in 1978 800,000 people experienced a first heart attack, while a further 200,000 had a second or third one. The odds are based on the population over the age of 15.

The Odds:	Heart attack	**158.0-1** against
	First heart attack	**197.0-1** against

Source: American Heart Association

WHAT ARE THE ODDS OF DYING ACCORDING TO AGE?

As the table shows, death, like taxes, is certain. However its cause changes radically as the years go by. Motor vehicle accidents account for one in five deaths among those under 14, while this grows to over one in three among 15 to 24-year olds. A disturbing new cause, suicide and murder, accounts for one in five in this age group. after the age of 45 various diseases are far and away the biggest killers.

The Odds: against

All ages	113.0-1
1-14	2,319.0-1
15-24	853.0-1
25-44	548.0-1
45-64	99.0-1
65-74	31.7-1
75 +	10.2-1

Source: National Safety Council, *Accident Facts*, 1979.

WHAT ARE THE ODDS OF DYING IN THE HOSPITAL OF SELECTED DISEASES OR ACCIDENTS?

The Commission on Professional and Hospital Activities is a research and education center which produces reports based on participating hospitals. During 1974-75 they monitored over 29 million patients in almost 1,900 hospitals. The odds which follow are compiled from that massively comprehensive survey.

Based on fatality rates per 10,000 patients admitted, these figures show that various forms of malignant cancer are far and away the most dangerous diseases. That won't come as a big surprise to most readers. But the report does contain some eyeopeners. Out of every 10,000 patients admitted with a fracture of the upper leg, 633 die. Ulcers don't just give rise to jokes, they kill—as do alcohol and even hemorrhoids.

We didn't list all the diseases; that would nearly be a book in its own right. We listed all those that account for very large numbers of patients as well as some that we're all familiar with.

The Odds: against

Malignant neoplasm of ill-defined and secondary sites	3.0-1
Malignant neoplasm of bronchus and lung	3.1-1
Miscellaneous cerebrovascular lesion with paralysis	3.7-1
Acute myocardial infarction	4.0-1
Miscellaneous cerebrovascular diseases	4.3-1
Malignant neoplasm of ovary	5.0-1
Malignant neoplasm of large intestine except appendix and rectum	5.6-1
Perinatal conditions	5.7-1
Heart failure	5.7-1
Miscellaneous diseases of liver	6.3-1
Intracranial injury except concussion	6.4-1
Lymphatic and hematopoietic neoplasm except Hodgkin's disease and leukemia	6.5-1
Laennec's cirrhosis	6.8-1
Malignant neoplasm of kidney and other urinary organs	6.9-1
Pulmonary embolism	7.3-1
Single preterm hospital newborn except that by c-section	7.4-1
Arterial embolism and thrombosis, gangrene	8.3-1
Malignant neoplasm of rectum and rectosigmoid junction	8.8-1
Arrhythmia and slowed conduction	10.9-1
Miscellaneous lung and pleural diseases	11.2-1
Malignant neoplasm of breast	14.2-1

The Odds: against

Fracture of upper end of femur	14.3-1
Pneumonia	14.6-1
Miscellaneous ischemic heart disease	18.1-1
Other peptic ulcer, complicated	25.1-1
Miscellaneous hypertensive disease	56.5-1
Diabetes mellitus without complication; chemical diabetes	58.2-1
Diverticular disease	79.6-1
Bronchitis, chronic and unspecified	103-1
Influenza	103-1
Rheumatoid arthritis	103-1
Special admissions and examinations without complaint or reported diagnosis	107-1
Benign prostatic hypertrophy	131-1
Diseases of gallbladder	148-1
Obesity of nonendocrine origin	153-1
Phlebitis and thrombophlebitis	160-1
Gastric ulcer without complication	160-1
Alcoholic addiction	166-1
Alcoholic mental disorder	166-1
Asthma	178-1
Angina pectoris	226-1
Diseases of thyroid gland	243-1
Bronchitis, acute	262-1
Osteoarthritis	269-1
Other peptic ulcer without complication	269-1
Fracture of tibia and fibula	269-1
Single term hospital newborn delivered by c-section	293-1
Concussion	293-1
Intestinal infectious diseases	322-1
Gastritis and duodenitis	322-1
Hernia of abdominal cavity except inguinal without complication	332-1
Cystitis	416-1
Facial bone fracture	416-1
Appendectomy	434-1
Acute URI except streptococcal	555-1
Fracture of radius and ulna	666-1
Varicose veins of leg	666-1
Single term hospital newborn except that by c-section	832-1
Senile cataract	908-1
Inguinal hernia without complication	999-1
Tooth extraction	1,249-1
Hemorrhoids	1,249-1

The Odds: against

Acute appendicitis without peritonitis	1,428-1
Diseases of ovary, fallopian tube, and parametrium, except endometriosis	1,666-1
Low cervical c-section	1,666-1
Dilation and curettage of uterus	4,999-1
Delivery complicated by umbilical cord complications	4,999-1
Normal delivery	9,999-1

Source: *Hospital Mortality*, Commission on Professional and Hospital Activities, Ann Arbor, Michigan 1977, pp. 277-310.

WHAT ARE THE ODDS BY DISEASE OR ACCIDENT OF DYING AT DIFFERENT AGES?

After the age of 25 cancer becomes a bigger killer than the car (16% versus 12% among those 25 to 44 years of age). Between the ages of 45 and 64 heart disease and cancer are nearly equal and account for very nearly two thirds of all deaths. After the age of 65 strokes become a major cause of death.

The Odds: against

All Ages	Total	Male	Female
All causes	113-1	100-1	129-1
Heart disease	300-1	265-1	344-1
Cancer	559-1	499-1	630-1
Stroke (cerebrovascular disease)	1,188-1	1,360-1	1,062-1
Accidents	2,095-1	1,463-1	3,558-1
Motor vehicle	4,366-1	2,940-1	8,129-1
Falls	15,624-1	14,492-1	16,948-1
Drowning	30,302-1	17,543-1	99,999-1
Fires, burns	34,482-1	27,026-1	45,454-1
Poisons (solid, liquid)	62,499-1	52,631-1	83,332-1
Pneumonia	4,328-1	3,875-1	4,877-1
Diabetes mellitus	6,578-1	7,751-1	5,746-1
Cirrhosis of liver	6,992-1	5,207-1	10,416-1
Arteriosclerosis	7,518-1	9,008-1	6,493-1
Suicide	7,518-1	4,974-1	14,705-1
Homicide	10,869-1	6,848-1	23,809-1
Emphysema	13,157-1	8,332-1	29,411-1

The Odds: against

Age 1 to 14	Total	Male	Female
Accidents	5,050-1	3,936-1	7,142-1
Motor vehicle	11,110-1	9,258-1	14,084-1
Drowning	27,777-1	18,518-1	55,555-1
Fires, burns	41,666-1	38,461-1	45,454-1
Firearms	124,999-1	83,332-1	333,332-1
Cancer	20,407-1	17,543-1	24,999-1
Congenital anomalies	27,777-1	27,777-1	28,570-1
Homicide	62,499-1	58,823-1	66,666-1
Pneumonia	71,428-1	66,666-1	71,428-1
Heart disease	90,908-1	90,908-1	90,908-1
Meningitis	166,666-1	142,856-1	199,999-1
Age 15 to 24			
Accidents	1,599-1	1,025-1	3,689-1
Motor vehicle	2,267-1	1,494-1	4,738-1
Drowning	19,230-1	10,525-1	99,999-1
Poison (solid, liquid)	58,823-1	41,666-1	99,999-1
Firearms	62,499-1	34,482-1	249,999-1
Suicide	7,352-1	4,586-1	18,867-1
Homicide	7,873-1	5,154-1	16,948-1
Cancer	15,384-1	12,345-1	20,407-1
Heart disease	39,999-1	132,257-1	52,631-1
Age 25 to 44			
Accidents	2,380-1		
Motor vehicle	4,347-1		
Drowning	33,332-1		
Poison (solid, liquid)	49,999-1		
Fires, burns	49,999-1		
Falls	49,999-1		
Other	9,999-1		
Cancer	3,332-1		
Heart disease	3,999-1		
Age 45 to 64			
Heart disease	248-1		
Cancer	329-1		
Stroke*	1,922-1		
Accidents	2,272-1		
Motor vehicle	5,555-1		
Falls	19,999-1		
Fires, burns	24,999-1		
Drowning	49,999-1		
Surg. complications	49,999-1		
Other	7,691-1		
Cirrhosis of liver	2,563-1		
Suicide	5,262-1		

*Cerebrovascular disease

The Odds: against

Age 65 to 74	**Total**
Heart disease	79-1
Cancer	125-1
Stroke*	384-1
Diabetes mellitus	1,514-1
Accidents	1,612-1
Motor vehicle	4,761-1
Falls	7,142-1
Fires, burns	16,666-1
Surg. complications	19,999-1
Ingestion of good, object	33,332-1
Other	7,691-1
Pneumonia	1,753-1
Cirrhosis of liver	2,325-1

Age 75 and over	
Heart disease	23-1
Stroke*	75-1
Cancer	76-1
Pneumonia	291-1
Arteriosclerosis	375-1
Accidents	587-1
Falls	1,148-1
Motor vehicle	8,332-1
Surg. complications	8,332-1
Fires, burns	9,090-1
Ingestion of food, object	12,499-1
Other	4,544-1
Diabetes mellitus	636-1
Emphysema	1,448-1

*Cerebrovascular disease

Source: *Accident Facts*, National Safety Council, Chicago, 1979, pp. 8-9.

WHAT ARE THE ODDS ON DYING ACCIDENTALLY FROM VARIOUS CAUSES?

In 1978 104,500 people died accidentally, an increase of 1% over 1977. Motor vehicle accidents accounted for just under 50% of this total.

The Odds: against

All accidents	2,087-1
Motor vehicle accidents	4,236-1
Falls	15,872-1
Drowning	31,249-1
Deaths associated with fire	34,482-1
Poisoning by solids/liquids	62,499-1
Suffocation	76,922-1
Firearms	124,999-1
Poisoning by gases and vapors	124,999-1
All others	13,512-1

Source: National Safety Council, *Accident Facts*, 1979.

WHAT ARE THE ODDS ON DYING ACCIDENTALLY BY STATE?

The odds increase drastically as you head west. New Jersey, New York, Rhode Island, Maryland and Washington, D.C. all have accidental death rates of around 30 people per 100,000 population. By way of contrast the figure for New Mexico is 81, for Nevada 84.4 and for Arizona and Montana 74. These are all states that rank high on motor fatalities.

The Odds: against

Total U.S.	2,087-1
Alabama	1,554-1
Alaska	——
Arizona	1,554-1
Arkansas	1,785-1
California	——
Colorado	2,052-1
Connecticut	——
Delaware	2,958-1
Dist. of Col.	3,066-1
Florida	1,886-1
Georgia	1,772-1
Hawaii	2,769-1
Idaho	1,591-1

The Odds: against

Illinois	2,631-1
Indiana	——
Iowa	2,104-1
Kansas	2,015-1
Kentucky	1,941-1
Louisiana	1,591-1
Maine	1,956-1
Maryland	3,011-1
Massachusetts	2,716-1
Michigan	2,617-1
Minnesota	——
Mississippi	1,507-1
Missouri	1,930-1
Montana	1,350-1
Nebraska	2,122-1
Nevada	1,184-1
New Hampshire	3,194-1
New Jersey	3,066-1
New Mexico	1,234-1
New York	3,508-1
North Carolina	1,714-1
North Dakota	1,574-1
Ohio	2,610-1
Oklahoma	1,708-1
Oregon	1,697-1
Pennsylvania	2,617-1
Rhode Island	3,412-1
South Carolina	1,557-1
South Dakota	1,741-1
Tennessee	1,723-1
Texas	1,785-1
Utah	1,788-1
Vermont	2,011-1
Virginia	2,131-1
Washington	1,907-1
West Virginia	——
Wisconsin	2,325-1
Wyoming	1,519-1
Puerto Rico	3,744-1
Virgin Islands	2,154-1

Source: National Safety Council, *Accident Facts,* 1979.

WHAT ARE THE ODDS BY CERTAIN CITIES OF DYING ACCIDENTALLY?

The National Safety Council has analyzed registrars of vital statistics for selected cities and these are shown below. Surprisingly, among this sample you're less apt to die accidentally in a major city than in a smaller one. On the average 25.9 people per 100,000 died accidentally in 1978 in the 10 cities with over 500,000 population, the figure for the next largest city was 35.0 while that for those with 200,000 to 350,000 citizens was 37.1. In cities below 200,000 the accidental death rate drops back to an average of 32.3.

The most fatal accident prone city on the list? It's New Orleans with 66.1 fatalities per hundred thousand, over three times the New York City number. Chicago has a tough reputation. It's undeserved. Based on these figures it has the lowest rate in our sample: just 17.4 fatal accidents per 100,000 residents.

The Odds: against

500,000 + Population (10 cities)	3,860-1
Baltimore	3,716-1
Chicago	5,746-1
Cleveland	1,995-1
Dallas	2,182-1
Detroit	2,923-1
New Orleans	1,512-1
New York	4,974-1
Phoenix	2,057-1
St. Louis	2,420-1
San Diego	2,363-1
Washington, D.C.	3,952-1
350,000-500,000 (7 cities)	2,856-1
Cincinnati	2,386-1
Denver	2,456-1
Kansas City, Mo.	2,197-1
Long Beach, Ca.	2,426-1
Omaha	4,366-1
Portland	3,267-1
Seattle	2,673-1
Toledo	2,761-1
200,000-350,000 (7 cities)	2,694-1
Birmingham	1,674-1
Norfolk	2,702-1
Oakland	2,432-1
Rochester	3,558-1
St. Paul	4,236-1

The Odds: against

Tampa	3,596-1
Tucson	2,432-1
100,000-200,000 (13 cities)	3,095-1
Allentown	4,853-1
Bridgeport	3,321-1
Fremont, Ca.	2,792-1
Hartford, Ct.	3,845-1
Independence, Mo.	4,131-1
Kansas City, Ka.	2,122-1
New Haven	4,097-1
Peoria	3,400-1
Roanoke	3,471-1
Rockford	3,545-1
Spokane	2,518-1
Springfield	2,403-1
Waterbury	3,343-1

Source: National Safety Council, *Accident Facts*, 1979.

WHAT ARE THE ODDS BY AGE AND SEX OF HAVING ARTHRITIS?

Arthritis is not only painful, it's prevalent. Thirty-one out of every 100 women suffer before they're 65. Half of all women do thereafter.

The Odds: against unless noted

	Men	Women
17-44	29.8-1	16.3-1
45-64	4.1-1	2.2-1
65 +	1.9-1	Even

Source: Department of Health, Education, and Welfare, *Prevalence of Chronic Skin and Musculoskeletal Conditions, United States—1976*, Series 10, No. 124, No. (PHS) 79-1552, 1978.

WHAT ARE THE ODDS BASED ON VARIOUS PERSONAL TRAITS THAT YOU'LL END UP WEARING DENTURES?

Denture wearing goes up as education and income go down. Women outnumber men and whites have a higher incidence than other races.

The Odds: against unless noted

Sex
Women	3.5-1
Men	4.1-1

Age
Under 30	44.4-1
30-39	3.8-1
40-49	1.9-1
50-59	1.2-1
60 +	1.3-1 for

Race
White	3.8-1
Black	5.0-1
Oriental	8.5-1
Other	11.0-1

Income
Under $6,000	1.7-1
$6,000-$9,999	3.5-1
$10,000-$14,999	4.9-1
$15,000-$19,999	5.2-1
$20,000 +	4.9-1

Education
Grade school only	1.5-1
Graduated from high school	2.8-1
Graduated from college	6.7-1

Source: *Prosthodontic Care: Number and Types of Denture Wearers*, American Dental Association, Bureau of Economic Research & Statistics, 1976.

WHAT ARE THE ODDS ON BEING A SCHIZOPHRENIC?

Two million people have been or would be diagnosed as schizophrenic.

The Odds: **89.4-1** against

Source: The President's Commission on Mental Health

WHAT ARE THE ODDS ON BEING MENTALLY RETARDED?

Six million Americans had some form of mental retardation in 1976.

The Odds: **34.2-1** against

Source: Ibid.

WHAT ARE THE ODDS ON BEING IN A MENTAL HOSPITAL, BASED ON SEX AND MARITAL STATUS?

In general women are far more prone to be institutionalized than men. The exception is among the formerly married where men have lower odds. Those who've never married are most apt to end up in a mental institution. It is probable that not having someone at home to care for them is a major contributing factor.

The Odds: against

Population 17 +	163.0-1
Men 17 +	230.0-1
Men married	728.0-1
Men formerly married	88.0-1
Men never married	58.0-1
Women 17 +	133.0-1
Women married	529.0-1
Women formerly married	97.0-1
Women never married	53.0-1

Source: Ibid.

WHAT ARE THE ODDS OF COMMITTING SUICIDE, BY AGE?

Our highest suicide age group is the 75 to 84-year-olds, at 62% above the national overall average.

The Odds: against

Overall	7,999-1
1-14	249,999-1
15-24	8,546-1
25-34	6,288-1
35-44	6,134-1
45-54	5,207-1
55-64	4,999-1
65-74	5,127-1
75-84	4,807-1
85 +	5,290-1

Source: Department of Health, Education and Welfare, *Final Mortality Statistics 1976*, National Center For Health Statistics, (PHS) 78-1120, Vol. 26, No. 12, Supplement 2, March 30, 1978.

Chapter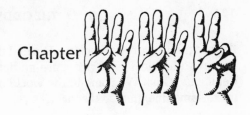

HOOKERS
Alcohol, Tobacco and Drugs

Any way you cut it we're a nation of addicts. Seven out of 10 of us drink, over six out of 10 have tried hash or marijuana by age 25, and over one out of three smokes.

Hookers analyzes the odds on these three forms of addiction in detail. Their use is by no means universal. Liquor never passes the lips of one out of four whites, four out of 10 blacks. Nonetheless they are common in our society.

What's more the character of our society is directly related to the nature of our addictions. Alcohol and drug use increase with education, while smoking declines.

Education isn't the only determinant. Alcohol use – and abuse – is tied to sex (male), religion (Catholic), age (youth) and geography (eastern). Alcoholism, with 10 million problem drinkers, is a serious national problem.

The odds show the situation is not apt to remedy itself in a hurry. Almost half of all teenagers drink every month. Of these, 7% had nine or more drinks during one bout. Of all teenagers who drink one out of two has been at the wheel of a car when he was drunk; one reason why alcohol figures in one out of every two accidents.

If alcohol is a pervasive addiction tobacco is a perverse one. Almost one out of four men and well over one out of three women smoke. Yet most don't want to. More than eight out of 10 have tried to kick the habit, yet only about one in five succeeds.

Finally we look at hard drugs in detail. The increase in their use has been meteoric in the past 10 years, so much so that among 18 to 25-year-olds experimentation and use of marijuana and/or hashish is greater than regular cigarette smoking. We end the chapter on a trip to nowhere: heroin addiction. There are more addicts than the entire population of Columbus, Ohio. Welcome to the world of trips; you, like Columbus, will make some unexpected discoveries.

WHAT ARE THE ODDS ON GETTING TANKED TODAY?

The average car consumes 706 gallons of gas per year, which works out to 13.9 gallons per week. The average tank capacity is 14 gallons. So ordinarily we all get tanked once a week.

The Odds: **6-1** against

Source: Data from Department of Energy.

WHAT ARE THE ODDS OF BEING HIGH?

Other than astronauts and test pilots no one flies higher than Concorde passengers. During the first two years of service, 144,000 flew the London/New York route, 107,000 the Paris/New York route. The airlines involved don't have an accurate count of American passengers versus other nationals. An educated guess is that 50% of these passengers are Americans and that the average passenger has flown the aircraft three times.

The Odds: **5,224-1** against

Source: Data from British Airways and Air France.

WHAT ARE THE ODDS ON NEVER TOUCHING ALCOHOL, BY RACE?

Over four out of 10 blacks never drink compared to one in four whites.

The Odds:	against
White	3.0-1
Black	1.4-1
Spanish-American	2.2-1
American Indian	1.7-1
Oriental	1.9-1

Source: Department of Health, Education, and Welfare, National Institute on Alcohol Abuse and Alcoholism, *Third Special Report to the U.S. Congress on Alcohol and Health, Technical Support Document,* Washington, D.C., June 1978.

WHAT ARE THE ODDS BY SEX ON BEING A CERTAIN TYPE OF DRINKER?

Men are far more apt to drink and to drink heavily than women.

The Odds: against

Type of Drinker	Men	Women
Abstain	2.8-1	1.6-1
Light	2.0-1	1.3-1
Moderate	3.2-1	5.7-1
Heavy	4.6-1	32.3-1

Source: Department of Health, Education and Welfare, National Institute on Alcohol Abuse and Alcoholism, *Contract (ADM) 281-76-0020,* July, 1977.

WHAT ARE THE ODDS ON DRINKING ALCOHOLIC BEVERAGES?

Today 69% of all Americans drink and the rate is growing all the time (up by 5% in the last 10 years). Drinking is least prevalent in the South (45% of those in the region abstain) and among those who only went to grade school (59% abstain). Catholics are big on booze: 83% indulge.

The Odds: for unless noted

National	2.2-1
Sex	
Men	2.8-1
Women	1.8-1
Education	
College background	3.8-1
High school	2.3-1
Grade school	1.4-1 against
Age	
18-29 years	4.0-1
30-49 years	2.9-1
50 years & older	1.3-1
Region	
East	3.0-1
Midwest	2.6-1
South	1.2-1
West	3.0-1
Religion	
Protestants	1.4-1
Catholics	4.9-1

Source: The Gallup Poll, Princeton, New Jersey, August 19, 1979.

WHAT ARE THE ODDS ON THE SEX OF AN ALCOHOLIC?

Of the nation's 10 million problem drinkers 2,500,000 are women.

The Odds: Men   **3.0-1** for
          Women   **3.0-1** against

Source: Department of Health, Education and Welfare, Public Health Service, Alcohol, Drug Abuse and Mental Health Administration, National Institute on Alcohol Abuse and Alcoholism, *Third Special Report to the U.S. Congress on Alcohol and Health*, June 1978.

WHAT ARE THE ODDS ON A SUICIDE BEING AN ALCOHOLIC?

The suicide rate among alcoholics is 58 times that of non-alcoholics.

The Odds: **2.2-1** against

Source: National Council on Alcoholism, Inc., *Facts on Alcoholism*, New York, 1974.

WHAT ARE THE ODDS ON HAVING ALCOHOL RELATED PROBLEMS OR BEING A CHRONIC ALCOHOLIC?

In 1977 there were 10,000,000 Americans in this category. The odds were figured by comparing this number to all those over the age of 15.

The Odds: **14.9-1** against

Source: President's Commission on Mental Health.

IF YOU HAVE AN ALCOHOL PROBLEM WHAT ARE THE ODDS ON RECEIVING HELP?

One million Americans were undergoing treatment in 1977.

The Odds: **9.0-1** against

Source: President's Commission on Mental Health.

WHAT ARE THE ODDS BY GRADE IN SCHOOL THAT A MALE OR FEMALE WILL BE A PROBLEM DRINKER?

By the end of the seventh grade 60.3% of all boys and 47.4% of all girls drink. This skyrockets to 88.5% of boys, 79.0% of girls by the end of their junior year in high school. As shown below the odds on being a problem drinker are alarmingly short.

The Odds: against

School Grade	Men	Women
7	19.0-1	21.7-1
8	5.4-1	10.0-1
9	3.8-1	5.2-1
10	2.7-1	4.3-1
11	2.0-1	4.0-1
12	1.5-1	3.9-1

Source: Department of Health, Education and Welfare, Public Health Service, Alcohol, Drug Abuse & Mental Health Administration, National Institute on Alcohol Abuse & Alcoholism, *Third Special Report to the U.S. Congress on Alcohol and Health*, June 1978, p. 45.

AMONG THE 47% OF HIGH SCHOOLERS WHO DRANK IN THE LAST MONTH WHAT ARE THE VARIOUS ODDS OF BEHAVIOR?

The Odds: against unless noted

Most alcohol drunk in one day during past week
9 or more drinks	6.1-1
4-8 drinks	2.4-1
1-3 drinks	1.6-1

Frequency of alcohol consumption
2 or more days per week	2.8-1
1 day a week	4.3-1
Once every 2 weeks	3.2-1
Less than once every 2 weeks	3.0-1

How many times driver of car when "really pretty drunk"
3 or more times	3.0-1
At least once	Even

Source: Department of Transportation, National Highway Traffic Safety Administration, Washington, D.C., *How to Talk to Your Teenager About Drinking and Driving.*

WHAT ARE THE ODDS ON ALCOHOL BEING A FACTOR IN A FATAL ACCIDENT ON THE ROADS?

The Odds: Even

Source: National Council on Alcoholism, Inc., *Facts on Alcoholism*, New York.

WHAT ARE THE ODDS OF SMOKING CIGARETTES, ACCORDING TO CERTAIN CHARACTERISTICS?

Between 1969 and January 1978 when the Gallup Organization last answered this question the overall percentage of the population that smokes declined by 4%. Nonetheless today 39% of all men and 34% of all women smoke. There are some other fairly marked differences between population groups.

Those who rate themselves as political independents smoke more than Democrats (44% versus 35%) and far more than Republicans (30%). Catholics outpuff Protestants hardily. Conversely, college graduates smoke less than those without a formal education.

The Odds: against

National	1.8-1	**Income**		
		$20,000 & over	1.7-1	
Sex		$15,000-$19,999	2.0-1	
Men	1.6-1	$10,000-$14,999	1.7-1	
Women	1.9-1	$ 7,000-$ 9,999	1.6-1	
		$ 5,000-$ 6,999	1.9-1	
Race		$ 3,000-$ 4,999	1.6-1	
White	1.7-1	Under $3,000	1.8-1	
Non-white	1.8-1			
		Politics		
Education		Republican	2.3-1	
College	2.3-1	Democrat	1.8-1	
High school	1.3-1	Independent	1.3-1	
Grade school	2.6-1			
		Religion		
Region		Protestant	1.8-1	
East	1.9-1	Catholic	1.5-1	
Midwest	1.6-1			
South	1.7-1	**Occupation**		
West	1.7-1	Professional & business	1.3-1	
		Clerical & sales	Even	
Age		Manual workers	1.4-1	
Total under 30	1.4-1	Non-labor force	2.4-1	
18-24 years	1.4-1			
25-29 years	1.4-1	**City Size**		
30-49 years	1.4-1	1,000,000 & over	1.8-1	
50 & older	2.7-1	500,000-999,999	1.6-1	
		50,000-499,000	1.8-1	
		2,500- 49,999	1.6-1	
		Under 2,500, rural	1.8-1	

Source: Gallup Poll, June, 1978.

WHAT ARE THE ODDS THAT A SMOKER WOULD LIKE TO KICK THE HABIT?

Over all about two out of three smokers would like to give it up.

The Odds: Men **1.6-1** for
Women **2.3-1** for

Source: Gallup Poll, June 1978.

WHAT ARE THE ODDS THAT A SMOKER WILL TRY TO QUIT?

Nationally 84% of all current smokers have attempted to stop at least once.

The Odds: Men **5.2-1** for
Women **4.9-1** for

Source: Gallup Poll, August 19-22, 1977.

HOW LONG DID PRESENT SMOKERS WHO TRIED TO QUIT SUCCEED?

Twenty-five percent of all men and 24% of all women couldn't last a week.

The Odds: against unless noted

	Men	Women	
Less than a week	3.8-1 for	3.2-1 for	
Less than a month	Even	1.2-1	
Less than 6 months	2.4-1	2.7-1	
1 year less than 2 years	9.0-1	9.0-1	

Source: Gallup Poll, August 19-22, 1977.

WHAT ARE THE ODDS THAT IF YOU SERIOUSLY TRY TO STOP SMOKING YOU'LL SUCCEED?

In 1978 17 million tried to stop. Only three and a half million succeeded (some of these will, of course, relapse).

The Odds: **3.8-1** against

Source: Data from National Center for Health Statistics.

WHAT ARE THE ODDS ON A CHILD FIRST TRYING MARIJUANA, BY SCHOOL GRADE?

First, for many parents, the (relatively) good news: 40.8% of those graduating from high school have never tried marijuana. Now the bad news: over one in four (28.2%) use the drug by the end of their freshman year in high school.

The Odds: against

6th grade	57.8-1
7th-8th grade	7.3-1
9th grade	5.9-1
10th grade	5.9-1
11th grade	8.3-1
12th grade	16.9-1
Never used	1.5-1

Source: Department of Health, Education, and Welfare, National Institute on Drug Abuse, No. (ADM) 79-878, *Highlights from Drugs and the Class of '78*, p. 34.

WHAT ARE THE ODDS ON HAVING USED CERTAIN DRUGS IN THE PAST MONTH, BY CURRENT AGE?

Just over one in four people aged 18-25 use marijuana monthly or more often, with one in 25 using cocaine this frequently.

The Odds: against

	12 to 17	18 to 25	26 +
Marijuana and/or hashish	5.2-1	2.6-1	30.2-1
Inhalants	142.0-1	N.A.	N.A.
Hallucinogens	61.5-1	49.0-1	N.A.
Cocaine	124.0-1	26.0-1	N.A.
Heroin	N.A.	N.A.	N.A.
Other opiates	166.0-1	99.0-1	N.A.
Stimulants*	75.9-1	39.0-1	166.0-1
Sedatives*	124.0-1	34.7-1	N.A.
Tranquilizers*	142.0-1	40.7-1	N.A.

*Non-medical use

N.A. = less than 0.5%

Source: *National Survey on Drug Abuse: 1977*, Vol. I, "Main Findings."

**WHAT ARE THE ODDS OF YOUTHS WITH CHOSEN CHARACTERIS-
TICS USING MARIJUANA AND/OR HASHISH, EVER AND RECENTLY?**

Whether you're describing those who have tried or who have recently
used these drugs the classic profile is the same: a white, 17-year-old
male living in a big city out west.

The Odds: against

Age	Ever used	In past month	
12-13	11.5-1	5.2-1	☞
14-15	2.4-1	2.4-1	
16-17	1.1-1	5.7-1	
Sex			
Men	2.0-1	4.3-1	
Women	3.3-1	6.7-1	
Race			
White	2.4-1	4.9-1	
Non-white	2.8-1	7.3-1	
Region			
Northeast	1.9-1	3.8-1	
North Central	2.4-1	4.3-1	
South	4.3-1	13.3-1	
West	1.8-1	3.5-1	
City size			
Large metropolitan	1.7-1	3.5-1	
Other metropolitan	2.6-1	5.3-1	
Non-metropolitan	4.6-1	9.0-1	

Source: *National Survey on Drug Abuse: 1977*, Vol. I, "Main Findings", p. 42.

WHAT ARE THE ODDS OF BEING A HEROIN ADDICT?

The 1977 report of the President's Commission on Mental Health cites
500,000 heroin addicts. The odds are based on the population over the
age of 15.

The Odds: **308-1** against ☞

Source: President's Commission on Mental Health.

WHAT ARE THE ODDS ON HAVING TRIED CERTAIN DRUGS, BY CURRENT AGE?

Almost one out of three 12 to 17-year-olds have tried marijuana or hashish. This jumps to six out of 10 by age 25. In addition, about one in five have tried one or more of all the following by this age: hallucinogens, cocaine, stimulants and sedatives.

The Odds: against unless noted

	12 to 17	18 to 25	26+
Marijuana and/or hashish	2.5-1	1.5-1 for	5.5-1
Inhalants	10.0-1	7.9-1	54.6-1
Hallucinogens	20.7-1	4.1-1	37.5-1
Cocaine	24.0-1	4.2-1	37.5-1
Heroin	89.9-1	26.8-1	124.0-1
Other opiates	15.4-1	6.4-1	34.7-1
Stimulants*	18.2-1	3.7-1	20.3-1
Sedatives*	31.3-1	4.4-1	34.7-1
Tranquilizers*	25.3-1	6.5-1	37.5-1

*Non-medical use

Source: Department of Health, Education and Welfare, Public Health Service, Alcohol, Drug Abuse and Mental Health Administration, *National Survey on Drug Abuse: 1977*, Vol. I, "Main Findings", pp. 19, 23, 25.

WHAT ARE THE ODDS ON VARIOUS DRUGS BEING "FAIRLY" OR "VERY" EASY FOR HIGH SCHOOL STUDENTS TO OBTAIN?

Most students in the class of 1978 reported that marijuana was easy to get (87.8% said so).

The Odds:

Marijuana	7.2-1 for
Tranquilizers	1.8-1 for
Amphetamines	1.4-1 for
Barbiturates	Even
Cocaine	1.6-1 against
LSD	2.1-1 against
Heroin	5.1-1 against

Source: Department of Health, Education and Welfare, National Institute on Drug Abuse, No. (ADM), 79-878, *Highlights from Drugs & The Class of '78*, p. 55.

WHAT ARE THE ODDS OF 18 TO 25-YEAR-OLDS WITH CHOSEN CHARACTERISTICS USING MARIJUANA AND/OR HASHISH, EVER AND RECENTLY?

Six out of 10 in this group have tried these drugs and usage is growing, so that about one out of three in the 18 to 21 age group has smoked in the past month.

The Odds: against unless noted

Age	Ever used	In past month
18-21	1.4-1 for	2.2-1
22-25	1.6-1 for	3.2-1
Sex		
Men	1.8-1 for	1.9-1
Women	1.2-1 for	3.8-1
Race		
White	1.6-1 for	2.6-1
Non-white	1.2-1 for	3.2-1
Region		
Northeast	1.9-1 for	1.9-1
North Central	1.6-1 for	2.3-1
South	Even	4.9-1
West	2.0-1 for	1.9-1
City Size		
Large metropolitan	1.7-1 for	2.2-1
Other metropolitan	1.8-1 for	2.3-1
Non-metropolitan	1.1-1	4.6-1
Education		
Not high school graduate	1.1-1 for	3.5-1
High school graduate	1.5-1 for	2.4-1
College drop-out	2.0-1 for	2.1-1
Now in college	1.8-1 for	2.3-1
College graduate	1.4-1 for	3.5-1

Source: *National Survey on Drug Abuse: 1977*, Vol. I, "Main Findings."

CONSUMING PASSIONS
What We Eat/What We Use

In the section which follows we examine our country in one of its world-famed roles—as the ultimate consumer society. In some ways consumption has gotten out of hand as shown by the odds which follow. For example, one out of four Americans overeats. We can't tell you whether or not other family members nag the fatties among us. But the odds tell us the opportunity is there because normally families share the dinner table. That is, they do when they're not going out to eat, which is often and primarily for a change of pace.

Whether we're out or at home we down soft drinks and coffee in vast amounts, so much that it's a wonder only one out of eight suffer persistent indigestion. Our stomachs may be tranquil but our heads are a different matter. Twenty percent of adults account for a whopping 75% of all headache remedy consumption.

From food and medication we turn our attention to hygiene and beauty. If you're a pessimist you'll find figures to gloom over; 10% of us never brush our teeth (let alone after every meal), one quarter of all males never use deodorant, and at least one third of blondes are probably not natural. Optimists will say the majority of adults are in great shape; with one third of all women going to the beauty parlor weekly, at least they try!

WHAT ARE THE ODDS ON CHICKENING OUT?

According to the main office of Kentucky Fried Chicken, on the average day 2,460,000 Americans eat at their outlets.

The Odds: **88.1-1** against

Source: Data from Kentucky Fried Chicken.

WHAT ARE THE ODDS ON OVEREATING?

One out of four Americans is overweight. A far greater number have occasional overweight concerns. Proof? Various research studies show that anywhere from 40% to 60% of the adult population attempt to diet during any given year.

You may be somewhat surprised to learn that we actually consume less calories than we did in years gone by. In 1900 the average American consumed 3,550 calories per day. Today the figure is about 3,160. That's a drop of 390 calories per day, 142,350 calories per year. Put another way, based on our average calorie intake today it's as though we built a 45 day fast into each year.

Based on this reduction alone we should be a nation of skinnys. The reality is something else, as you'll note in any crowd. The reasons are many. First and foremost we lead far easier, more sedentary lives than our ancestors. We may deposit fewer calories in our "body bank" but we're also making fewer withdrawals due to less exercise. Secondly, the percentage of animal derived proteins we eat is going up and these foods are denser in the ratio of calories to weight. That makes them less healthy. Thirdly, we're on a sweet-tooth kick, substituting less complex carbohydrates (such as sugar) for more complex ones (such as starch).

The Odds: **3-1** against

Sources: T. Scala, *U.S. Consumption Patterns Serve as Indices for Dietary Product Design*, Emeryville, California, Shaklee Corporation, March, 1978 and "Weight Watcher's Magazine Survey", *Housewares*, February, 1974.

WHAT ARE THE ODDS ON HAVING TROUBLE BECAUSE YOU ATE THE WHOLE THING?

Fairly slight, since 51.9% of adults never use indigestion aids or upset stomach remedies and only 12.5% of us use them once a week or more.

The Odds:

Use of indigestion aid in any given week   **7.0-1** against

Source: Target Group Index

WHAT ARE THE ODDS ON A FAMILY EATING DINNER TOGETHER?

Seventy-five percent of all families enjoy the evening meal together. In almost three out of 10 homes (28%) there's an extra family member heard from—the televison set is on during the meal.

The Odds:

Total	3.0-1 for
If wife employed full time	1.7-1 for
If wife employed part time	2.2-1 for
If wife unemployed	3.8-1 for

Source: *The Gallup Report,* Vol. 3, No. 2, 1979.

WHAT ARE THE ODDS OF BEING OUT TO LUNCH?

Across the nation the average work week is now 37 hours and 30 minutes. On average lunch is half an hour. We allow that those on Madison Avenue do better, but to compensate, some of us work right through lunch.

The Odds: **15.1-1** against

Source: Department of Labor, Bureau of Labor Statistics.

WHAT ARE THE ODDS OF OWNING A RECREATIONAL VEHICLE?

Four and a half million American families (8%) owned a recreational vehicle in mid-1978.

The Odds: **11.5-1** against

Source: Recreational Vehicle Industry Association, 1978.

WHAT ARE THE ODDS OF OWNING A SECOND OR VACATION HOME?

That house in the country, at the beach, or elsewhere is still very much a dream.

The Odds: **23.4-1** against

Source: U.S. Bureau of the Census

WHAT ARE THE ODDS YOU'LL BUY A NEWSPAPER EVERY DAY?

Compared to some other countries we are not as well read. Evidently the evening news on T.V. is all we want to know.

The Odds: **24.1** against

Source: Heron House, *Book of Numbers,* New York: A & W Publishers, Inc., 1978.

WHAT ARE THE ODDS ON EVEN HIS/HER BEST FRIEND WON'T TELL HIM OR HER?

It all depends on whether the problem is body odor, bad breath or dandruff. According to a Roper Poll conducted in 1977 91% of adults bathe or shower in any given 24 hours. While Target Group Index reports that 72.1% of men and 82.9% of women use a shampoo and a hefty 11.2% do so daily, 11.6% of the population never use toothpaste and 34.9% never use a mouthwash. Could these be the 51.9% of adults who use breath-fresheners?

The Odds: for

An adult bathing daily	10.1-1
A man using a deodorant daily	2.6-1
A woman using a deodorant daily	4.8-1
A woman using shampoo	7.7-1
An adult using toothpaste	7.6-1
An adult using a breath-freshener	1.1-1

Source: *Progressive Grocer's Guide To: 73 Most-Used Health & Beauty Aids Products*, Target Group Index, Spring 1978.

WHAT ARE THE ODDS ON "DOES SHE OR DOESN'T SHE"?

According to the Target Group Index report which rates Americans on their use of various products 28.3% of women use hair coloring products at home. An undertermined small added number only have their hair dyed professionally. Hair dying is most common among lower income, widowed, divorced or separated women in the Southeast.

The Odds: **2.5-1** against

Source: Target Group Index

WHAT ARE THE ODDS ON HEARING "NOT TONIGHT, I HAVE A HEADACHE"?

Six and a half percent of adults use headache remedies and pain relievers daily while another 9.7% use these remedies at least two or three times a week. These 19.7% of total users account for almost three quarters of the total consumption (73.4%). The odds are based on the daily users.

The Odds: **14.4-1** against

Source: Target Group Index

WHAT ARE THE ODDS ON GETTING TELLTALE LIPSTICK ON YOUR COLLAR?

Excellent: 69% of women use it daily.

The Odds: **2.2-1** for

Source: Ibid.

WHAT ARE THE ODDS THAT YOU'LL "MAKE UP" TODAY?

About 32% of women use make-up daily while 50.3% never touch the stuff.

The Odds: **2.2-1** against

Source: *Progressive Grocer Guide To: 73 Most-Used Health & Beauty Aids Products,* Target Group Index, 1978, p. 149.

WHAT ARE THE ODDS OF A WOMAN REMOVING EXCESS HAIR?

Virtually all women do so (98.5%). Of this total 75.7% sometimes use a safety razor, 44.9% sometimes use an electric razor and 18.8% use only depilatories. The normal areas of the body de-fuzzed are the legs (86.7%) and the underarms (84.4%). That's not surprising. What is is the 8.3% who remove hair from their face regularly.

The Odds:

On excess hair removal	**65.7-1** for	
On excess facial hair removal	**11.0-1** against	

Source: *Seventeen's Personal Daintiness Survey,* ETJ Computer Service, 1975, p. 9.

WHAT ARE THE ODDS THAT A MAN HAS SHAVED TODAY?

Men over the age of 24 shave an average of six times a week while those under 24 only shave an average of four times per week.

The Odds:	Men over 24	**5.7-1** for	
	Men under 24	**1.3-1** for	

Source: Gillette

WHAT ARE THE ODDS ON GOING TO A BEAUTY SALON AT LEAST ONCE A MONTH?

Sixty-three percent of women patronize a beauty salon at least once a month, while 35% of all women go weekly.

The Odds: **1.7-1** for

Source: Data from The Market For Haircutting and Styling, FIND/SVP, 1977.

WHAT ARE THE ODDS WOMEN WILL USE MASCARA?

More women use mascara than eyeshadow although not by much.

The Odds: **Even**

Source: Heron House, *The Book of Numbers*, New York: A & W Publishers Inc., 1978.

WHAT ARE THE ODDS WOMEN USE FACE POWDER?

Since of all forms of make-up this is the least used we assume that ladies don't go to the powder room to powder as often as we think.

The Odds: **2.0-1** against

Source: Ibid.

WHAT ARE THE ODDS WOMEN WILL USE PERFUME OR TOILET WATER?

American women use more perfume/toilet water than lipstick. We have no way of knowing if most of it is purchased by men for their ladies.

The Odds: **5.7-1** for

Source: Ibid.

WHAT ARE THE ODDS WOMEN WILL USE ROUGE OR BLUSHER?

Maybe we just have a lot of naturally rosy cheeked females.

The Odds: **1.3-1** against

Source: Ibid.

WHAT ARE THE ODDS ON YOUR FAVORITE WAY OF SPENDING AN EVENING?

No surprises here; T.V. is our favorite pastime of these five activities. There is no one thing in particular that is an all American favorite, except maybe watching commercials.

The Odds: against

Watching television	1.2-1
Reading	6.1-1
Sport (as participant)	19.0-1
House and vehicle maintenance	32.3-1
Parties, entertaining, visiting friends	5.3-1

Source: The Gallup Organization, Inc.

WHAT ARE THE ODDS YOU WILL GO SHOPPING IN THE NEXT 48 HOURS FOR SOMETHING OTHER THAN FOOD?

In most cases there will be only one trip made (47%) although 18% will make three or more trips.

The Odds:

Men	**Even**
Women	**1.1-1** against

Age	
18-24 years	**Even**
25-34 years	**1.1-1** for
35-49 years	**1.2-1** for
50 years & older	**1.6-1** against

Source: *The Gallup Report*, The Gallup Organization, June, 1978.

WHAT ARE THE ODDS THAT YOUR TRIP WAS WINDOW SHOPPING?

We all know women do more window shopping than men but surprisingly it is not that much different. Weekend shoppers are only slightly more disposed toward browsing.

The Odds:

Men	**4.0-1** against
Women	**3.4-1** against

Source: Ibid.

WHAT ARE THE ODDS YOU ENJOY SHOPPING?

The Odds: against

	Women	Men
Great deal	1.2-1	5.7-1
Somewhat	2.1-1	1.4-1
Not at all	3.6-1	1.4-1

Source: Ibid.

WHAT ARE THE ODDS YOU HAVE MADE A CATALOGUE PURCHASE IN THE LAST TWO WEEKS?

About one in eight people have except those aged 50 years and older, who seem to be somewhat less likely to. Regionally, catalogue shopping is least popular in the West and most popular in the Midwest.

The Odds: **6.1-1** against

Source: Ibid.

WHAT ARE THE ODDS YOU WENT (NON-FOOD) SHOPPING BY YOUR-SELF?

Men are more likely to shop alone, although of those that shop with someone it is most likely to be with their spouse (34%). Twenty-two percent of women shop with children rather than their spouse (16%).

The Odds: Men   **1.1-1** against
                       Women   **2.1-1** against

Source: *The Gallup Report*, The Gallup Organization, Inc., July, 1978.

WHAT ARE THE ODDS YOUR SHOPPING TRIP (NON-FOOD) WAS MADE TO A MALL?

It's not surprising that our Main Streets are dying, more than half of those who have shopped in the last two days went to a mall. Seventy-one percent of the mall shoppers went to an enclosed mall.

The Odds: **1.2-1** for

Source: Ibid.

WHAT ARE THE ODDS OF SELDOM OR NEVER EATING CERTAIN FOODS?

Only two really show as being seldom or never eaten, skim milk and artificially sweetened beverages. Others with a good chance of seldom being eaten are fish and shellfish, cereals, candy and salty snacks to name a few.

The Odds: against unless noted

	Above poverty level		Below poverty level	
	White	Black	White	Black
Whole milk	3.9-1	3.6-1	5.1-1	3.6-1
Skim milk	5.4-1 for	9.2-1 for	7.0-1 for	6.1-1 for
Meat & poultry	249.0-1	166.0-1	42.5-1	199.0-1
Fish & shellfish	1.2-1	1.5-1	Even	1.5-1
Eggs	4.6-1	5.5-1	4.6-1	4.8-1
Cheese	4.8-1	2.0-1	2.6-1	1.5-1
Legumes, seeds & nuts	2.0-1	2.1-1	3.0-1	4.3-1
Fruits & vegetables (all kinds)	332.0-1	499.0-1	82.3-1	61.5-1
Fruits & vegetables (rich in Vitamin A)	1.8-1	3.8-1	1.4-1	3.6-1
Fruits & vegetables (rich in Vitamin C)	6.6-1	6.1-1	4.2-1	6.0-1
Bread	99.0-1	142.0-1	57.8-1	142.0-1
Cereals	1.6-1	1.4-1	1.7-1	1.9-1
Fats & oils	11.5-1	6.8-1	6.6-1	4.6-1
Desserts	7.3-1	6.1-1	5.5-1	6.0-1
Candy	1.6-1	1.5-1	4.2-1	2.3-1
Sweetened non-alcoholic beverages	2.5-1	6.4-1	2.7-1	6.5-1
Artificially sweetened non-alcoholic beverages	5.9-1 for	9.2-1 for	9.9-1 for	15.9-1 for
Coffee & tea	3.3-1	2.1-1	2.7-1	1.7-1
Salty snacks	1.7-1	2.0-1	1.6-1	2.4-1

Source: Department of Health, Education, and Welfare, Vital Health Statistics, *Food Consumption Profiles of White and Black Persons Aged 1-74 Years: United States, 1971-74*, Series 11, No. 210.

WHAT ARE THE ODDS OF HAVING CERTAIN FOODS AT LEAST ONCE A DAY?

It is no surprise that all of us eat meat, vegetables and bread at least once a day. Oddly not all of use milk, coffee or tea. But we all are definitely not fish eaters.

The Odds: against unless noted

| | Above poverty level | | Below poverty level | |
	White	Black	White	Black
Whole milk	1.4-1 for	1.1-1	1.6-1 for	1.1-1 for
Skim milk	10.1-1	34.7-1	16.2-1	26.0-1
Meat & poultry	6.1-1 for	5.4-1 for	2.7-1 for	4.5-1 for
Fish & shellfish	99.0-1	110.0-1	99.0-1	89.9-1
Eggs	6.1-1	3.7-1	3.4-1	3.1-1
Cheese	6.7-1	15.9-1	9.0-1	19.0-1
Legumes, seeds & nuts	11.5-1	9.8-1	4.2-1	6.2-1
Fruits & vegetables all	12.5-1 for	5.3-1 for	4.8-1 for	3.6-1 for
Fruits & vegetables (rich in Vitamin A)	24.0-1	8.7-1	28.4-1	14.1-1
Fruits & vegetables (rich in Vitamin C)	1.6-1	1.7-1	2.4-1	2.0-1
Bread	7.3-1 for	6.4-1 for	6.2-1 for	6.9-1 for
Cereals	4.9-1	7.2-1	4.3-1	4.4-1
Fats & oils	2.7-1 for	1.3-1 for	1.7-1 for	1.1-1 for
Desserts	1.4-1	1.7-1	1.8-1	1.6-1
Candy	4.9-1	3.1-1	3.8-1	2.2-1
Sweetened non-alcoholic beverages	1.9-1	1.1-1	2.0-1	1.1-1
Artificially sweetened non-alcoholic beverages	24.0-1	30.3-1	34.7-1	51.6-1
Coffee & tea	1.7-1 for	1.3-1	1.3-1 for	1.7-1
Salty snacks	8.1-1	5.7-1	9.9-1	3.7-1

Source: Ibid.

WHAT ARE THE ODDS YOU'LL SPEND MONEY ON CERTAIN ITEMS ON YOUR NEXT TRIP TO THE SUPERMARKET?

The American diet has come up for attack more than once. Our figures indicate that you are likely to spend money on calories and cholesterol. Most of your money is spent on meat. That may well be due to the sheer price of meats and cold cuts which demand almost one dollar in four. Fresh fruit is followed by dairy foods, which means that some of the myths about living on convenience food are not reflected by supermarket buying, although the fifth most likely candidate, bread, cakes, and that favorite fattener, potato chips, account for about 7.5% of supermarket spending. So the odds indicate that the majority of your money is spent on widening the waistline!

The Odds: against

Baked goods, snacks	12.2-1
Dairy products	12.2-1
Frozen foods	14.9-1
Beer and wine	14.2-1
Cereals and rice	51.6-1
Candy and chewing gum	70.4-1
Canned fruit & vegetables	23.4-1
Canned juice and drinks	89.9-1
Canned meat and poultry	70.4-1
Canned seafood	110.0-1
Canned soups	124.0-1
Fresh fruit & vegetables	6.4-1
Soft drinks	29.3-1
Fresh meat & processed	3.2-1
Fresh fish	110.0-1
Fresh poultry	33.5-1
Macaroni, spaghetti, noodles	199.0-1

Chapter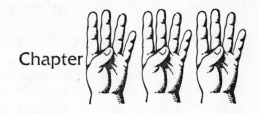

ON THE ROAD
Motorcycles, Cars and Buses

If you were looking for the perfect prescription for a traffic fatality it would go like this. The driver would be a well educated young man (under 25) who now works in New Mexico as an executive with a labor union. It's Saturday night, he's on his way home from work and he's had too much to drink. He isn't wearing a seat belt (86% of us ignore them). He is driving a sub-compact. The weather's clear. He's going to hit something head on and the odds are quite clear as to who will be killed: he will.

We examine our depressing penchant for maiming and killing ourselves on the nation's highways and byways (most deaths occur within five miles of the victim's home).

The damage is, quite simply, enormous. Over one in nine cars will be in a collision in any given year. More than half of all accidental fatalities are vehicle related.

If cars are dangerous, motorcycles are lethal. One motorcyclist is killed each year for every 1,251 on the road but the injuries are appalling. Mount a motor bike in Maryland, New York or Louisiana and your odds are well under one in 20 of surviving the year without an accident.

There is a more accident prone vehicle roaming our streets: the school bus. One out of seven is in an accident each year. The odds on a child being injured in such an accident are 871,999-1 nationally.

Other kinds of buses have excellent safety records. While mass transit advocates normally use fuel savings as their main argument, safety statistics are even more compelling. Your odds of being killed on a long distance bus are 21 million-1, on a local bus, 231 million-1. Those really are favorable odds!

WHAT ARE THE ODDS ON PUTTING IN A COLLISION CLAIM BY SIZE OF CAR?

Out of every 100 sub-compacts about 13 will be in some form of collision, based on data supplied by the Highway Loss Data Institute on 1978/79 models. The sub-compact's average repair claim was $1,129, very nearly the $1,131 for full-size cars.

The Odds: against

All 78-79 models	7.5-1
Sub-compact (wheelbase 101 inches or less)	6.6-1
Compact (101 to 111 inches)	7.7-1
Intermediate (111 to 120 inches)	8.9-1
Full-size (120 inches or more)	7.8-1

Source: Highway Loss Data Institute, *Automobile Insurance Losses—Collision Coverages Initial Results for 1979 Models*, p. 4.

WHAT ARE THE ODDS OF BEING THE DRIVER IN A COLLISION OR ACCIDENT WITH ANOTHER CAR?

A large minority of Americans, over four in 10, have been in a car crash. Myth notwithstanding men are almost twice as likely as women to have been involved in such an accident. At 58% male executives share with male union members the highest rate.

Education correlates directly with accidents. The more schooling you've had the more apt you are to have a car accident. The figures: 53% for those who've gone to college, 41% for those who only attended high school and 24% for those who only completed grade school.

Predictably about three out of four say it was the other driver's fault.

The Odds: **1.3-1** against

Source: *Los Angeles Times* Syndicate, The Roper Poll, 1977, p. 75.

WHAT ARE THE ODDS OF WEARING A SEAT BELT?

That seat belts save lives can be readily proven. See our odds on fatalities. Yet we continue to refuse to buckle up in massive numbers. Nationally only 14.1% of drivers use belts. The figure climbs to almost one in five (18.5%) on the West Coast, while in the Southwest just over half that number (9.8%) wear seat belts.

The Odds: against

Nationally	6.1-1	Southwest	9.2-1
New England	7.3-1	North Central	8.8-1
Mid Atlantic	7.5-1	West	4.4-1
Southeast	6.6-1		

Source: Department of Transportation, Opinion Research Corporation, National Highway Transportation Safety Administration, *Safety Belt Usage,* December 1978.

WHAT ARE THE ODDS OF WEARING A SEAT BELT BASED ON MAKE OF CAR DRIVEN?

Surprisingly foreign car owners, and notably Volkswagen Rabbit owners, are far more likely to wear seat belts than those in Detroit's products.

The Odds: against

VW Rabbit	1.9-1
Misc. foreign	3.1-1
Toyota	3.9-1
VW other	4.1-1
Datsun	4.9-1
Chrysler	6.4-1
AMC	7.5-1
GM	8.2-1
Ford	8.3-1

Source: Ibid.

WHAT ARE THE ODDS, BY SEVERITY OF INJURY, ON BEING HURT IN A CAR CRASH WITH AND WITHOUT SEAT BELTS?

The Department of Transportation commissioned a special study among more than 15,000 1973 and 1975 cars damaged seriously enough to warrant a tow truck.

The Odds: against unless noted

	No or Minor injury	Moderate injury	Serious injury	Fatal
No seat belt	5.2-1 for	7.3-1	30.3-1	124.0-1
Seat belt	12.7-1 for	16.2-1	75.9-1	499.0-1

Source: Department of Transportation, National Highway Safety Administration, (HS) 5-01255, Vol. 1, 1976.

WHAT ARE THE STATE BY STATE ODDS THAT IN A CAR CRASH WITH ANOTHER AUTOMOBILE THE OTHER CAR WILL BE INSURED?

We compared automobiles registered with those insured on a state by state basis. In six states other insured vehicles creep into these figures so you have the curious phenomenon of more cars insured then there are registered! This may also be the result of two policies being taken out on the same car within one year. We marked these states as virtually 100% auto insured.

Over and above these states those where insured autos are the vast majority include, in rank order, Wisconsin, Iowa, Missouri and Delaware. If you're driving from Delaware to D.C. your odds on being hit by an uninsured vehicle change radically. Over one out of three vehicles in the nation's capitol is uninsured: the worst record in the country.

The Odds: for

Alabama	2.3-1
Alaska	2.4-1
Arizona	10.2-1
Arkansas	14.5-1
California	4.2-1
Colorado	7.3-1
Connecticut	8.7-1
Delaware	100% insured
D.C.	1.7-1
Florida	4.6-1
Georgia	8.4-1
Hawaii	3.5-1
Idaho	3.7-1
Illinois	4.2-1
Indiana	7.0-1
Iowa	44.4-1
Kansas	100% insured
Kentucky	8.4-1
Louisiana	3.3-1
Maine	4.5-1
Maryland	13.3-1
Massachusetts	4.3-1
Michigan	8.9-1
Minnesota	100% insured
Mississippi	2.0-1

The Odds: for

Missouri	34.7-1
Montana	7.4-1
Nebraska	100% insured
Nevada	4.6-1
New Hampshire	7.6-1
New Jersey	7.6-1
New Mexico	5.0-1
New York	7.0-1
North Carolina	31.2-1
North Dakota	100% insured
Ohio	4.4-1
Oklahoma	17.8-1
Oregon	2.3-1
Pennsylvania	3.9-1
Rhode Island	3.1-1
South Carolina	22.2-1
South Dakota	100% insured
Tennessee	5.0-1
Texas	2.7-1
Utah	9.5-1
Vermont	5.2-1
Virginia	5.8-1
Washington	11.0-1
West Virginia	100% insured
Wisconsin	75.9-1
Wyoming	100% insured

Source: Insurance Information Institute, *Insurance Facts 1979*, New York, p. 31.

WHAT ARE THE ODDS ON FILING A CLAIM UNDER YOUR AUTO COLLISION COVERAGE BY CAR CATEGORY FOR 1979?

For every 100 sub-compact cars almost 14 had put in a claim. One would think that full-size cars would be least prone to damages. Curiously this isn't so. Intermediate models have consistently had the lowest incidence of collision claims.

The Odds: against

Sub-compact	6.2-1
Compact	7.6-1
Intermediate	9.4-1
Full size	8.0-1

Source: Highway Loss Data Institute, *1979 Models*, 1979.

WHAT ARE THE ODDS ON PUTTING IN AN INSURANCE LOSS CLAIM FOR CERTAIN 1979 CAR MODELS?

These certainly vary drastically by model. For every 100 years of insurance (an industry term that can be equated with 100 cars insured for one year) 18.8 Toyota Celicas will register a claim. That's almost double the rate registered by 4-door Chevettes (9.8). The average loss payment per Celica claim is $1,265. At $256 more than the average claim on a two-door Cadillac de Ville, the Celica has the highest average damage claim among the entire sample.

If you're in the market for a new car these odds should enter into your considerations.

The Odds: against

Sub-Compacts		
		6.2-1
Pontiac	Sunbird 2-Dr.	7.3-1
Chevrolet	Monza 2-Dr.	5.8-1
Chevrolet	Chevette 2-Dr.	8.1-1
Volkswagen	Rabbit 2-Dr.	7.4-1
Ford	Pinto 2-Dr.	6.4-1
Ford	Mustang 2-Dr.	6.3-1
Toyota	Corolla 2-Dr.	5.2-1
Mercury	Capri 2-Dr.	5.8-1
Toyota	Celica 2-Dr.	4.3-1
Chevrolet	Chevette 4-Dr.	9.2-1
Plymouth	Horizon 4-Dr.	8.8-1
Dodge	Omni 4-Dr.	7.1-1

The Odds: against

Compact		7.6-1
Chevrolet	Nova 2-Dr.	9.3-1
Ford	Granada 2-Dr.	8.5-1
Chevrolet	Malibu 2-Dr.	9.0-1
Oldsmobile	Cutlass 2-Dr.	8.0-1
Buick	Century/Regal 2-Dr.	7.1-1
Ford	Fairmont 2-Dr.	6.4-1
Chevrolet	Nova 4-Dr.	9.6-1
Chevrolet	Malibu 4-Dr.	12.0-1
Chevrolet	Malibu St. Wgn.	8.3-1
Ford	Fairmont St. Wgn.	8.6-1
Chevrolet	Monte Carlo Specialty	6.9-1
Pontiac	Grand Prix Specialty	7.9-1
Chevrolet	Camaro Specialty	5.8-1
Pontiac	Firebird Specialty	5.2-1
Intermediate		9.4-1
Oldsmobile	Delta 88 2-Dr.	12.7-1
Ford	LTD 2-Dr.	11.2-1
Oldsmobile	Delta 88 4-Dr.	13.3-1
Buick	LaSabre 4-Dr.	12.3-1
Ford	LTD 4-Dr.	11.7-1
Chevrolet	Caprice	10.4-1
Buick	Electra 4-Dr.	10.4-1
Chevrolet	Impala 4-Dr.	8.9-1
Pontiac	Bonneville 4-Dr.	10.0-1
Oldsmobile	Ninety Eight 4-Dr.	8.7-1
Ford	Thunderbird Specialty	7.5-1
Chrysler	Cordoba Specialty	6.9-1
Mercury	Cougar XR-7 Specialty	6.6-1
Full Size		8.0-1
Cadillac	DeVille 4-Dr.	8.7-1
Cadillac	DeVille 2-Dr.	8.3-1
Lincoln	Continental 4-Dr.	7.8-1
Lincoln	Mark V	6.9-1

Source: Highway Loss Data Institute, Automobile Insurance Losses, *Collision Coverages, Variations By Make and Series,* 1979.

WHAT ARE THE ODDS OF BEING INVOLVED IN A FATAL TRAFFIC ACCIDENT?

In 1977 there were 42,064 fatal traffic accidents involving 47,715 deaths. During the same year there were 13,790,000 driver licenses in force. Thus 3.05 out of 10,000 drivers were involved in a fatal accident.

The Odds: **3,178-1** against

Source: Department of Transportation, National Highway Traffic Safety Administration, National Center for Statistics and Analysis, *1977 F.A.R.S. Annual Report,* 1977, p. 2.

WHAT ARE THE ODDS THAT THE DRIVER IN A FATAL ACCIDENT WILL BE OF A CERTAIN AGE AND GENDER?

Almost one out of two drivers (48%) in accidents which kill are under 25. Up to the age of 65 very nearly three out of four such drivers are men.

The Odds: against

Age	Men	Women
25 or less	1.8-1	6.7-1
26-35	7.0-1	27.6-1
36-50	9.5-1	27.6-1
51-65	11.0-1	26.0-1
65 +	13.9-1	22.2-1

Source: Department of Transportation, National Highway Traffic Safety Administration, National Center for Statistics and Analysis, *1977 Fatal Accident Reporting System Annual Report,* Vol. II, 1977, p. 13.

WHAT ARE THE ODDS, BY STATE, ON BEING INVOLVED IN A FATAL ACCIDENT PER 5,000 MILES DRIVEN?

California leads the nation in fatal accidents with 4,351 in 1977, the latest year available for the Department of Transportation's National Highway Traffic Safety Administration. Texas is next, then New York. Yet this large number is chiefly a reflection of their large populations. They are far from the most dangerous states in which to drive. This dubious distinction goes to a grouping of our western states. New Mexico with 5.09 fatalities per 100 million miles driven heads the list, followed by Wyoming, Arizona, Nevada, Idaho and Montana.

The West's traditionally reckless attitude toward law and order may have something to do with its high fatality rate, as do the wide open spaces. Long stretches of empty road encourage speeding and speed kills.

This hypothesis is borne out by the safest states. Washington, D.C. leads this list with just 1.76 fatalities per 100 million miles, little more than a third of New Mexico's rate. The next safest are a group of fairly small, densely urbanized eastern states: New Jersey, Rhode Island, Connecticut, Maryland and Massachusetts, in that order.

One hundred million miles is such a large number it seems like an abstract. Five thousand isn't; many of us drive that in a year. On that basis it takes just 20,000 drivers to notch up 100 million miles. That's how we arrived at the odds.

The Odds: against

	Fatal accidents per 100 million vehicle miles travelled	Based on 5,000 miles per year
Alabama	3.40	5,881-1
Alaska	3.87	5,166-1
Arizona	4.41	4,523-1
Arkansas	3.23	6,190-1
California	2.88	6,943-1
Colorado	3.30	6,059-1
Connecticut	2.09	9,568-1
Delaware	2.51	7,967-1
D.C.	1.76	11,362-1
Florida	2.71	7,379-1
Georgia	2.76	7,245-1
Hawaii	2.89	6,919-1
Idaho	4.04	4,949-1
Illinois	2.92	6,848-1
Indiana	2.63	7,603-1
Iowa	2.60	7,691-1
Kansas	2.88	6,943-1
Kentucky	2.92	6,848-1
Louisiana	3.87	5,166-1
Maine	2.49	8,031-1
Maryland	2.20	9,089-1
Massachusetts	2.22	9,008-1
Michigan	2.65	7,546-1
Minnesota	2.59	7,721-1
Mississippi	3.24	6,171-1
Missouri	3.01	6,643-1
Montana	3.99	5,011-1

	Fatal accidents per 100 million vehicle miles travelled	Based on 5,000 miles per year
Nebraska	2.41	8,297-1
Nevada	4.16	4,806-1
New Hampshire	2.31	8,657-1
New Jersey	1.86	10,751-1
New Mexico	5.09	3,928-1
New York	3.08	6,492-1
North Carolina	3.12	6,409-1
North Dakota	3.16	6,328-1
Ohio	2.38	8,402-1
Oklahoma	2.90	6,895-1
Oregon	3.23	6,190-1
Pennsylvania	2.54	7,873-1
Rhode Island	1.93	10,361-1
South Carolina	3.52	5,680-1
South Dakota	3.18	6,288-1
Tennessee	2.97	6,733-1
Texas	3.25	6,152-1
Utah	3.39	5,898-1
Vermont	2.87	6,967-1
Virginia	2.54	7,873-1
Washington	2.91	6,871-1
West Virginia	3.91	5,114-1
Wisconsin	2.50	7,999-1
Wyoming	4.62	4,328-1
TOTAL	3.47	5,762-1

Source: Ibid., Vol. I, p. 11.

WHAT ARE THE ODDS OF HAVING A FATAL VEHICLE ACCIDENT ON A CERTAIN DAY OF THE WEEK?

Over half of all fatalities (54%) occur on Friday, Saturday and Sunday, while Monday, Tuesday and Wednesday each have very nearly the same accident rates.

The Odds: against

Saturday	3.8-1	Wednesday	8.1-1
Sunday	5.2-1	Thursday	7.3-1
Monday	8.1-1	Friday	4.9-1
Tuesday	8.1-1		

Source: Ibid.

WHAT ARE THE ODDS WHEN THE DRIVER IS KILLED THAT THE CAR WILL HAVE BEEN HIT AT VARIOUS IMPACT POINTS?

Just over half of all driver fatalities are the result of front end collisions.

The Odds: against

Head on	1.4-1
Left front	15.7-1
Right front	25.5-1
Left side	5.2-1
Right side	7.6-1
Rear	36.0-1
Top	9.9-1
Non-collision (Rollover, etc.)	12.5-1

Source: Department of Transportation, National Highway Traffic Safety Administration, National Center for Statistics and Analysis, *1978 Fatal Accident Reporting System Annual Report*, Vol. I, 1978, p. 26.

WHAT ARE THE ODDS ON WHO THE VICTIM WILL BE IN A TRAFFIC DEATH?

Over half of fatalities (55%) are drivers. Passengers are the next group at risk (27% of deaths), then pedestrians (16%).

The Odds: against unless noted

Situation of person when killed

Driver	1.2-1 for
Passengers	2.7-1
Other—Includes children standing, people on laps, in back of station wagon, etc.	16.5-1
Passengers front right	5.4-1
Passengers rear left	40.7-1
Passengers front middle	61.5-1
Passengers rear right	61.5-1
Pedestrians	5.1-1
Pedal cyclist	51.6-1

Source: Ibid.

WHAT ARE THE ODDS OF HAVING A FATAL VEHICLE ACCIDENT AT A CERTAIN TIME OF DAY?

Nearly half of all fatal accidents (4.3 out of 10) occur between 4:00 in the afternoon and midnight.

The Odds: against

Midnight-3:59 A.M.	4.0-1
4:00 A.M.-7:59 A.M.	11.5-1
8:00 A.M.-11:59 A.M.	9.0-1
Noon-3:59 P.M.	5.2-1
4:00 P.M.-7:59 P.M.	3.3-1
8:00 P.M.-11:59 P.M.	3.5-1

Source: Ibid.

WHAT ARE THE ODDS THAT A TRAFFIC FATALITY WILL TAKE PLACE UNDER GIVEN WEATHER CONDITIONS?

Fully 75% of all fatalities take place in clear conditions.

The Odds: against unless noted

Clear	3.0-1 for
Cloudy	7.3-1
Rain	9.0-1
Snow	49.0-1
Fog, smoke, wind, sand or dust	49.0-1
Sleet	99.0-1

Source: Ibid.

WHAT ARE THE ODDS BY STATE OF BEING HURT IN A MOTORCYCLE ACCIDENT?

In 1977 there were 171,252 reported motorcycle accidents out of 5,144,129 registered motorcycles. On a national basis, this comes out as 333 accidents per 10,000 registered motorcycles.

Curiously some of our most dangerous states for auto accidents are some of our safest states for motorcyles: Montana, North Dakota, Oregon and Pennsylvania, in that order. All are over twice as safe as the national average.

Conversely one of the states with the best auto safety records, Maryland, is the most dangerous state for motorcycling. Their accident rate is more than double the national average and six times that of Montana's per 1,000 registrations. For every 13 Free State motorcyclists there was one accident during 1977. The next most dangerous states, in order, are New York, Louisiana and Hawaii.

The Odds: against

	Accidents per 10,000 registrations
Alabama	22.5-1
Alaska	33.3-1
Arizona	18.1-1
Arkansas	21.2-1
California	22.5-1
Colorado	52.4-1
Connecticut	24.5-1
Delaware	18.0-1
Florida	21.3-1
Georgia	27.3-1
Hawaii	17.2-1
Idaho	62.5-1
Illinois	26.8-1
Indiana	28.1-1
Iowa	58.5-1
Kansas	41.9-1
Kentucky	32.4-1
Louisiana	16.7-1
Maine	33.1-1
Maryland	12.7-1
Massachusetts	23.3-1
Michigan	33.7-1
Minnesota	54.2-1
Mississippi	49.3-1
Missouri	29.0-1
Montana	89.1-1
Nebraska	37.0-1
Nevada	24.5-1
New Hampshire	49.5-1
New Jersey	17.3-1
New Mexico	24.1-1
New York	15.2-1
North Carolina	31.3-1
North Dakota	71.1-1
Ohio	28.8-1

The Odds: against

	Accidents per 10,000 registrations
Oklahoma	48.1-1
Oregon	70.1-1
Pennsylvania	65.1-1
Rhode Island	30.1-1
South Carolina	22.3-1
South Dakota	52.6-1
Tennessee	29.8-1
Texas	24.4-1
Utah	42.9-1
Vermont	53.6-1
Virginia	20.2-1
Washington	34.3-1
West Virginia	43.4-1
Wisconsin	38.6-1
Wyoming	45.8-1

WHAT ARE THE ODDS ON BEING KILLED ON A MOTORCYCLE?

In 1977 there were 4,111 motorcycle fatalities out of 5,144,129 registered motorcycles. Causing 7.99 deaths per 10,000 motorcycles, this form of transportation is as dangerous as many people claim.

The Odds: **1,250-1** against

WHAT ARE THE ODDS BY STATE ON YOUR CHILD BEING IN A SCHOOL BUS ACCIDENT OR BEING INJURED IN SUCH AN ACCIDENT?

As shown below the national odds per year are alarming.

These are National Safety Council 1978 estimates based on years of experience. Their figures add up to 58,000 annual school bus accidents out of a total of 380,000 school buses and 4,500 pupil injuries on the bus (defined as absent from school at least one day) out of 21,800,000 pupils transported daily.

We used the states' own reporting methods in compiling the odds below which are virtually all based on the 1977-78 school year. For all of the reasons listed the figures should be viewed critically. It is fair to say that parents in West Virginia and Maryland should have some qualms as they watch their kids climb aboard a bus. In each of these states the odds are less than 700-1 against an accident in any given year.

In terms of pupil injuries Washington, D.C. is the nation's worst performer, with one injury for every 84,000 pupils carried. Curiously, Maryland comes out on top in this category. So though their buses are in accidents they must be minor ones and their definition of an accident must be especially rigid. Other notable injury free states are Hawaii, Florida and Wyoming.

The Odds: against

	School bus being in an accident	Pupil being injured
U.S. (Total)	1,178-1	871,999-1
Alabama	3,049-1	1,773,873-1
Alaska	1,728-1	NR
Arizona	2,662-1	2,153,275-1
Arkansas	2,744-1	1,336,504-1
California	1,574-1	NR
Colorado	10,499-1	1,172,499-1
Connecticut	838-1	477,854-1
Delaware	1,478-1	456,029-1
Dist. of Col.	NR	83,999-1
Florida	1,549-1	2,973,951-1
Georgia	1,819-1	1,646,340-1
Hawaii	3,703-1	6,513,119-1
Idaho	1,997-1	344,482-1
Illinois	NR	NR
Indiana	1,532-1	870,229-1
Iowa	2,220-1	654,515-1
Kansas	3,106-1	NR
Kentucky	1,174-1	84,986,999-1
Louisiana	1,901-1	434,376-1
Maine	2,833-1	448,074-1
Maryland	672-1	8,123,710-1
Massachusetts	NR	NR
Michigan	898-1	NR

The Odds: against

	School bus being in an accident	Pupil being injured
Minnesota	9,374-1	NR
Mississippi	2,136-1	308,679-1
Missouri	1,611-1	49,025-1
Montana	10,709-1	0
Nebraska	3,453-1	NR
Nevada	3,139-1	360,760-1
New Hampshire	27,119-1	NR
New Jersey	2,886-1	549,295-1
New Mexico	1,315-1	985,897-1
New York	3,455-1	436,970-1
North Carolina	1,658-1	476,975-1
North Dakota	4,363-1	624,227-1
Ohio	902-1	1,253,082-1
Oklahoma	1,685-1	297,824-1
Oregon	1,761-1	897,927-1
Pennsylvania	2,577-1	712,383-1
Rhode Island	1,380-1	640,576-1
South Carolina	1,245-1	160,465-1
South Dakota	2,650-1	931,517-1
Tennessee	1,264-1	415,411-1
Texas	1,609-1	199,933-1
Utah	3,651-1	1,206,248-1
Vermont	1,909-1	1,499,259-1
Virginia	1,881-1	541,314-1
Washington	3,349-1	NR
West Virginia	590-1	1,126,600-1
Wisconsin	1,410-1	270,683-1
Wyoming	2,930-1	2,021,459-1

NR = Not Reported

Source: National Safety Council, *Accident Facts 1979*, Chicago, Illinois, 1979, p. 93.

WHAT ARE THE ODDS OF BEING KILLED IN AN INTERSTATE BUS ACCIDENT?

Here's the bad news. For the latest reporting year, 1977, these increased by 200%. Now for the good news. That increase was from two to six fatalities per 124,900,000 passengers carried.

The Odds: **20,816,665-1** against

Source: Data from American Bus Association.

WHAT ARE THE ODDS OF BEING KILLED IN A LOCAL BUS ACCIDENT?

In 1976, the latest reporting year, there were 11 passenger fatalities out of 4,168,000,000 individual passenger trips.

The Odds: **231,554,550-1** against

Source: Data from American Public Transportation Association.

Chapter

TAKEOFF/CAST OFF
Airplanes and Boats

In 1980 over 14 million boats will take to the water. By the most conservative estimates one out of five Americans over the age of 12 will go boating at least once.

If boating is big, flying is enormous. Industry projections indicate well over 400 million passengers will board United States regularly scheduled flights in 1980 alone.

In this chapter we examine how well man does in these two unnatural environments of water and sky. In both, mishaps normally occur not when nature strikes but when the captain goofs.

Coast Guard figures convincingly show that yachtsmen could cut way back on accidents with proper training. Curiously, the obvious correlation doesn't hold true for private pilots. Those with 100 to 300 hours of flying time are most accident prone. Here it's a clear case of a little experience being a dangerous thing.

There are other interesting contrasts. Weather in general, and air turbulence specifically, cause many aircraft injuries but relatively few fatalities. By way of contrast, boating fatalities occur when the seas are calm, the wind light and the weather clear.

Other surprises abound in the pages ahead. Rented boats account for a relatively insignificant number of deaths. Air taxis are far more dangerous than other private planes. You're four times more apt to die while landing than while taking off in a commercial plane. The size of your boat, its type, age, construction and engine all directly relate to the odds of taking (or missing) a fatal voyage.

If you're a white-knuckle flyer you'll be reassured at the very long odds of being killed on that next coast to coast flight. If, on the other hand, you fly every week, you may well pick up your calculator to figure out exactly how many hours you've flown thus far and how many you have "left."

If you're reading this while tied up to the dock you'll find the odds on a fatality occurring under this condition. Isn't it about time to check those lines?

WHAT ARE THE ODDS ON BEING INVOLVED IN A FATAL BOATING ACCIDENT?

In 1977 the Coast Guard estimated that there were 13,600,000 boats on the water in the United States. During that year there were 1,312 boating fatalities.

The Odds: **10,365-1** against

Source: Department of Transportation, *Coast Guard Boating Statistics,* 1977, p. 7.

IF YOU ARE INVOLVED IN A FATAL BOATING ACCIDENT WHAT ARE THE ODDS THAT THE BOAT WAS RENTED?

Of the 1,133 fatal accidents in one report just 72 boats were rented. Thus it's not the Sunday sailors that account for the bulk of serious accidents.

The Odds: **14.7-1** against

Source: Department of Transportation, *Coast Guard General Boating Statistics,* 1977, p. 23.

WHAT ARE THE ODDS OF BEING IN A FATAL BOATING ACCIDENT ON A CERTAIN DAY OF THE WEEK?

Predictably, about 40% of all fatal accidents occur on weekends.

The Odds: against

Friday	9.8-1
Saturday	2.4-1
Sunday	3.0-1
Monday	11.0-1
Tuesday	11.7-1
Wednesday	9.7-1
Thursday	8.6-1

Source: Department of Transportation, *Coast Guard General Boating Statistics,* 1977, p. 23.

WHAT ARE THE ODDS OF BEING INVOLVED IN A FATAL ACCIDENT AT A CERTAIN TIME OF DAY?

Over one out of four accidents occurs between lunch and 4:30 in the afternoon. While Coast Guard records do not correlate accidents with inebriation, discussions with them indicate that many of these fatalities are the result of alcohol impairing the judgement of a skipper.

The Odds: against

12:30 to 4:30 P.M.	2.0-1
4:30 to 8:30 P.M.	2.3-1
8:30 to midnight	9.7-1
Midnight to 4:30 A.M.	21.1-1
4:30 A.M. to 8:30 A.M.	19.0-1
8:30 A.M. to 12:30 P.M.	4.6-1

Source: Department of Transportation, *Coast Guard Boating Statistics*, 1977, p. 23.

WHAT ARE THE ODDS BY MONTH OF BEING INVOLVED IN A BOATING FATALITY?

Just over half of all fatal accidents take place during the summer months of May, June, July and August.

The Odds: against

January	29.5-1	July	5.7-1
February	27.5-1	August	8.7-1
March	11.7-1	September	11.3-1
April	10.7-1	October	12.4-1
May	6.1-1	November	25.8-1
June	6.3-1	December	21.2-1

Source: Department of Transportation, *Coast Guard Boating Statistics*, 1977, p. 23.

WHAT ARE THE ODDS, IF THERE IS A BOATING FATALITY, OF THE OPERATOR HAVING HAD FORMAL TRAINING?

In 81% of all boating fatalities the operator has had no formal training.

The Odds:

Formal training	**4.5-1 against**
None	**4.5-1 for**

Source: Department of Transportation, *Coast Guard Boating Statistics*, 1977, p. 20.

WHAT ARE THE ODDS ON HAVING A FATAL BOATING ACCIDENT, BY OPERATION BEING UNDERTAKEN WHEN FATALITY OCCURRED?

Well over half of all accidents occurred during some form of cruising. Yet you're not wholly without risk when your boat is moored or anchored. Almost 10% of fatalities occurred during these situations.

The Odds: against

Operation at time of accident

Cruising	1.4-1
Cruising, fishing	10.2-1
Cruising, hunting	45.8-1
Cruising, sailing	22.4-1
Maneuvering	39.3-1
Maneuvering, docking	1,169.0-1
Maneuvering, undocking	389.9-1
Maneuvering, mooring	584.0-1
Water skiing	28.2-1
Water skiing, maneuvering with skiier down	389.9-1
Racing	89.0-1
Towing	233.0-1
Being towed	1,169.0-1
Drifting	7.2-1
Drifting, fishing	8.5-1
Drifting, hunting	129.0-1
Drifting, diving or swimming	291.5-1
Drifting, fueling	584.0-1
At anchor	34.5-1
At anchor, fishing	28.2-1
At anchor, fueling	584.0-1
Tied to dock	57.5-1
Other	28.2-1

Source: Department of Transportation, *Coast Guard Boating Statistics*, 1977, p. 19.

WHAT ARE THE ODDS THAT THE OPERATOR OF A VESSEL ON WHICH A FATALITY OCCURS CONTRIBUTED TO THE DEATH?

In two out of three cases this is so.

The Odds: **2-1** for

Source: Coast Guard Boating Statistics

WHAT ARE THE ODDS THAT THE BOAT'S OPERATOR WILL BE A CERTAIN AGE WHEN A DEATH OCCURS?

The odds are better than even that the operator will be between 26 and 50 years old.

The Odds: against unless noted

Age of operator	
Under 12	218.8-1
12-18	10.6-1
19-25	4.0-1
26-50	Even
51 +	4.1-1

Source: Coast Guard Boating Statistics

WHAT ARE THE ODDS ON VARIOUS WATER AND WEATHER CONDITIONS WHEN A BOATING FATALITY OCCURS?

As the figures below show most fatalities occur in ideal water and weather conditions. Nature isn't the culprit; unseaworthy boats and inexperienced skippers are.

The Odds: against

Water Conditions	
Calm	1.1-1
Choppy	3.2-1
Rough	6.7-1
Very rough	8.6-1
Strong current	14.6-1
Wind	
None	4.3-1
Light	1.5-1
Moderate	3.8-1
Strong	5.6-1
Storm	19.9-1
Visibility	
Good	3.0-1
Fair	6.3-1
Poor	17.8-1
Dark	16.3-1

Source: Department of Transportation, *Coast Guard Boating Statistics,* 1977, p. 22.

WHAT ARE THE ODDS ON THE NUMBER OF PEOPLE ON BOARD WHEN A FATALITY OCCURS?

The odds are better than even that only one or two people will be on board.

The Odds: against

Number on Board	
1	4.6-1
2	1.7-1
3	3.4-1
4	6.9-1
5 +	9.0-1

Source: Department of Transportation, *Coast Guard Boating Statistics*, 1977, p. 21.

WHAT ARE THE ODDS ON VESSEL TYPE WHEN A BOATING FATALITY OCCURS?

As shown by the odds you can readily build a profile of the typical boat involved in a fatality. It's an open, fiberglass motorboat under 16 feet long and at least nine years old, with an outboard motor generating over 26 horsepower.

The Odds: against unless noted

Year Built	
1977	8.9-1
1976	6.6-1
1974-1975	5.4-1
1972-1973	7.0-1
1969-1971	4.7-1
1964-1968	6.0-1
Prior to 1964	4.9-1

Type Boat	
Open motorboat	1.3-1 for
Cabin motorboat	13.2-1
Auxiliary sailboat	31.8-1
Sailboat only	23.9-1
Rowboat	8.6-1
Canoe or kayak	7.8-1
Inflatable boat	24.2-1
Houseboat	228.6-1
Other	22.4-1

The Odds: against unless noted

Length

Less than 16 feet	1.1-1 for
16 feet to less than 26 feet	1.4-1
26 feet to less than 40 feet	17.1-1
40 feet to less than 65 feet	154.1-1

Hull material

Wood	6.3-1
Aluminum	1.8-1
Steel	76.3-1
Fiberglass	1.1-1
Rubber, vinyl, canvas	39.0-1
Other	359.7-1

Propulsion

Outboard	1.2-1 for
Inboard	9.0-1
Inboard diesel	92.4-1
Inboard-outboard	23.4-1
Jet	92.4-1
Sail	23.4-1
Manual (oars, paddle)	5.8-1
Other	79.1-1

Horsepower

No engine	2.1-1
10 hp or less	5.8-1
11-25 hp	9.3-1
26-75 hp	4.6-1
Over 75 hp	2.9-1

Source: Department of Transportation, *Coast Guard Boating Statistics*, 1977.

WHAT ARE THE ODDS ON BEING KILLED IN A CRASH BY A REGULARLY SCHEDULED COMMERCIAL PLANE?

Obviously these can vary enormously by year. One mid-air collision over San Diego or a disintegrating engine at O'Hare can seriously affect the statistics.

What's more there are all kinds of ways to compute these odds. The most basic is the number of passengers boarded versus the number of fatalities, but you can also increase it by passenger miles, hours in the air or takeoffs and landings (where most accidents occur).

We took the last five years' worth of statistics to present a more balanced picture.

The Odds: against

Based On

Number of passengers	603,930-1
Number of hours in the air	21,772-1
Number of takeoffs/landings	33,370-1
Number of passenger miles	8,959,000-1

Source: National Transportation Safety Board, *Annual Review of Accident Data 1974-1977, Preliminary Report,* 1978.

WHAT ARE THE ODDS ON HAVING AN ACCIDENT IN A NON-SCHEDULED AIRCRAFT BASED ON THE PILOT'S FLIGHT EXPERIENCE IN TERMS OF HOURS FLOWN?

Surprisingly there is no direct correlation between experience and safety as the following figures show.

The Odds: against

Pilot Total Time

0-25 hours	28.5-1	1,001-3,000 hours	3.8-1
26-50 hours	28.7-1	3,001-5,000 hours	9.8-1
51-100 hours	12.1-1	5,001-8,000 hours	15.0-1
101-300 hours	4.8-1	8,001-10,000 hours	44.6-1
301-500 hours	9.1-1	Over 10,000 hours	20.8-1
501-1,000 hours	6.6-1		

Source: National Transportation Safety Board, *Annual Review of Aircraft Accident Data, U.S. Air Carrier Operations,* 1977, p. 100.

WHAT ARE THE ODDS PER 1,000 HOURS FLOWN ON BEING IN AN ACCIDENT OR DYING IN THE CRASH OF A NON-SCHEDULED FLIGHT?

Of all such accidents only about one in six results in death.

The Odds:	Of being in an accident	**900,800-1** against
	Of dying	**5,498,500-1** against

Source: National Transportation Safety Board, *Annual Review of Aircraft Accident Data,* 1977.

WHAT ARE THE ODDS ON BEING IN AN ACCIDENT IN VARIOUS TYPES OF NON-SCHEDULED AIRCRAFT USES?

If you happen to be a crop duster or fire fighter they're depressingly good: almost 17 accidents for every 100,000 hours flown. A pleasure flyer's rate—over 18 accidents—is even worse.

Surprisingly corporate jets and business charters are much safer than air taxis.

The Odds per 100 thousand hours: against

Kind of flying	
Instructional	10,615-1
Pleasure	5,419-1
Business	25,705-1
Corporate	66,224-1
Aerial application	5,892-1
Air taxi	19,379-1

Source: National Transportation Safety Board, *Annual Review of Aircraft Accident Data 1977.*

WHAT ARE THE ODDS OF AN AIRPLANE ACCIDENT BEING RELATED TO ONE OF THE TEN MOST COMMONLY CITED CAUSES?

As shown by the figures below it's pilot error, not weather, that causes the great bulk of accidents.

The Odds: against

Pilot—Failed to obtain/maintain flying speed	2.6-1
Weather—Low ceiling	3.9-1
Pilot—Continued VFR flight into adverse weather conditions	5.2-1
Weather-Fog	5.9-1
Pilot—Spatial disorientation	6.1-1
Pilot—Inadequate preflight preparation or planning	6.2-1
Terrain—High obstructions	7.2-1
Pilot—Improper inflight decisions or planning	8.1-1
Miscellaneous—Unwarranted low flying	8.5-1
Miscellaneous—Undetermined	11.9-1

Source: National Transportation Safety Board, *Annual Review of Aircraft Data, U.S. General Aviation,* 1977, pp. 13-14.

WHAT ARE THE ODDS ON BEING IN A PLANE CRASH, BASED ON THE PHASE OF FLIGHT IN A NON-REGULARLY SCHEDULED PLANE?

If you're a white-knuckle flyer in a private plane or air taxi, the time to get out the worry beads is as you land: 43% of all accidents occur at this point.

The Odds: against

Plane stationary	116.6-1
Taxiing	24.8-1
Takeoff	3.8-1
In flight	20.2-1
Landing	1.3-1

Source: National Transportation Safety Board, *Annual Review of Aircraft Acci dent Data, U.S. General Aviation,* 1977, p. 9.

WHAT ARE THE ODDS ON WHAT PORTION OF THE FLIGHT IS UN DER WAY WHEN A FATAL ACCIDENT OCCURS ON A REGULARLY SCHEDULED AIRCRAFT?

Landings account for over four out of 10 fatal commercial airline ac cidents. Takeoff is four times safer.

The Odds: against

Static	10.9-1	Inflight	2.0-1
Taxiing	22.8-1	Landing	1.4-1
Takeoff	9.1-1		

Source: Ibid.

WHAT ARE THE ODDS ON THE PRIMARY CAUSE OF NON-FATAL ACCIDENTS IN REGULARLY SCHEDULED AIR CARRIERS?

An analysis of the last 10 years shows that one or more passengers is injured in air turbulence, at just under one third of all causes (32%).

The Odds: against

Turbulence	2.1-1
Collision with wires, trees, tower, etc.	10.4-1
Collision with ground/water	15.4-1
Engine failure	16.5-1
Gear collapsed	19.0-1
Fire or explosion on ground	26.8-1
Aircraft collision both on ground	30.2-1

Source: National Transportation Safety Board, *Accident Data, 1968-1977,* p. 18, 19.

WHAT ARE THE ODDS ON THE PRIMARY CAUSE OF FATAL ACCIDENTS IN REGULARLY SCHEDULED AIR CARRIERS?

While turbulence is a major non-fatal accident cause, a straightforward collision is most lethal.

The Odds: against

Primary cause of fatal accident	
Collision with ground/water	2.4-1
Collision with wires, trees, tower, etc.	6.1-1
Midair collisions	7.8-1
Undershoot runway	16.8-1
Turbulence	16.8-1

Source: National Transportation Safety Board, *Annual Review of Aircraft Accident Data, U.S. Air Carrier Operations, 1977.*

WHAT ARE THE ODDS OF DIFFERING ELEMENTS CONTRIBUTING TO A FATAL COMMERCIAL ACCIDENT?

In nearly two out of three cases (62.5%) pilot error is a contributing factor, while weather plays a factor in close to half of all fatal accidents (45.3%).

The Odds: against unless noted

Causes or related factors			
Pilot error	1.7-1 for	Airports/facilities	20.3-1
Weather	1.2-1	Powerplant	20.3-1
Other personnel	1.4-1	Landing gear	31.2-1
Systems	9.6-1	Instruments/equipment	31.2-1
Airframe	15.1-1		

Source: Ibid.

WHAT ARE THE ODDS ON BEING IN AN ACCIDENT OR DYING IN THE CRASH OF A NON-REGULARLY SCHEDULED FLIGHT?

In 1977 there were 4,286 aircraft accidents involving 4,337 aircraft. Of the 8,625 people in these planes 1,425 died (16.5% of those aboard).

During this year these non-scheduled aircraft logged a total of 38,600,000 hours. The odds are based on the number of hours flown.

The Odds:		
	Of being in an accident	**8,899-1** against
	Of being killed	**27,087-1** against

Source: National Transportation Safety Board, *Annual Review of Aircraft Accident Data, U.S. General Aviation, 1977,* p. 25.

Chapter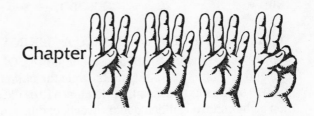

SHEER CATASTROPHE!
Fires, Floods and Other Disasters

It's small wonder that smoke alarms have become popular enough to warrant television advertising. If past years are anything to go by, in 1980 we'll experience well over 800,000 residential fires causing over three billion dollars' worth of damage. That's an average of over $3,000 per home fire, which puts the cost of an alarm and, for that matter, adequate fire insurance, into proper perspective.

In *Sheer Catastrophe!* we examine lightning and fires (the two are notably connected out west), earthquakes, hurricanes, floods and tornadoes. The odds on being hit by lightning are low. Also, it's one of nature's few threats which the potential victim can easily avoid. You probably know it's wise to dismount or discontinue that golf game at the sight of the first flash. But did you know that your telephone could be a lethal instrument?

Virtually all weather related phenomena are determined by where you live, so where possible, we give figures by states. Tornadoes are a good example. Alaskans can relax on this score – none there – while it's a threat that should make Texans tense. They have one nearly every second day.

Alaskans and Californians can worry about earthquakes instead. Both states are prone to this terrifying phenomenon. Each has had a killer in recent years. Our breakdown shows where they have hit lately, but it isn't necessarily indicative of what might occur. You can win a lot of money by betting on the area where many scientists think our worst earthquake occurred: then a territory, today it's part of Missouri. Yet this state (at two per year) is low on our current earthquake list.

Here then are the odds on the acts of the gods with, as in the case of fires, a little help from their friend: man's carelessness.

WHAT ARE THE ODDS ON HAVING A FIRE IN YOUR HOME?

In 1979 there were 797,000 residential fires. Of this total 85% were in one and two family dwelling units while the balance were in apartments (92,000 of the total). In combination these fires did an astounding two billion eight hundred million dollars worth of damage, an average of $3,574 for every fire.

The Odds: Of having a domestic fire per year **92.0-1** against

Source: Data from National Fire Prevention and Control Administration.

WHAT ARE THE ODDS OF DYING IN A FIRE?

In 1977 a total of 6,600 people died from injuries which were the direct result of conflagrations.

The Odds: **32,730-1** against

Source: Data from National Safety Council.

IF YOU DO HAVE A FIRE WHAT ARE THE ODDS OF IT BEING TOTALLY DESTRUCTIVE?

According to the National Fire Prevention and Control Administration, of all domestic fires attended by fire departments 13% of one and two family dwelling fires spread to an entire floor or the entire structure. The comparative figure for apartment fires is 4.8%.

The Odds:

Fires in one & two family dwelling units	**6.7-1** against
Fires in apartments	**19.8-1** against

Source: Data from National Fire Prevention and Control Administration.

WHAT ARE THE ODDS ON BEING STRUCK BY LIGHTNING?

The National Safety Council reported 116 lightning deaths in 1977. They don't supply data on those struck but not killed, yet other studies, such as one provided by the Lightning Protection Institute, show that there are over two injuries reported for every death. In addition, many injuries go unreported. By extrapolation you can arrive at a rough minimum number of deaths and injuries for that year: 360.

This only takes into account direct deaths and injuries from the bolt itself as opposed to the after effects of fires, electrocution from downed cables and injury from falling trees. If these figures were added in they'd increase the risk factor considerably. Lightning is the single largest creator of forest fires in the western United States.

The danger of being hit by lightning goes way up for certain categories. Over one out of three occurrences is on a beach, pier or actually in the water, while horseback riding and golf are similarly hazardous undertakings in a thunderstorm. Reasonably enough, over 50% of injuries are in the 10 to 35-year-old age bracket and fully 83% of fatalities are male.

You're a lot safer in your own home but not completely. Charges transmitted via an appliance are a menace, so stay out of your kitchen and laundry room and don't watch TV. Above all don't use your telephone. It can be a killer; studies show it accounts for about 1% of all deaths.

The danger is geographically selective. While annual figures show lightning strikes humans in all lower 48 states, Alaska and Hawaii rarely report injuries or deaths from this cause. The Gulf Coast (notably around Tampa) and Colorado's mountains are high risk areas, as are the places along our major waterways and their tributaries: the Mississippi, Ohio and Hudson.

Big cities are safer but not risk free. In June of 1973 two children were killed and seven injured by a single bolt of lightning in the heart of New York's Central Park. The overall odds are for any given year.

The Odds: **606,944-1** against

Sources: National Safety Council, *Accident Facts,* Chicago, Illinois, 1979, p. 12 and Lightning Protection Institute, *Lightning Protection for Home, Farm and Family,* Harvard, Illinois, pp. 6, 7, 12, 13.

WHAT ARE THE ODDS OF BEING KILLED BY A CATACLYSM?

According to the National Safety Council the total deaths from this cataclysm category in 1976 was 212.

The Odds: **980,000-1** against

Source: National Safety Council, *Accident Facts,* Chicago, Illinois, 1978, p. 12.

WHAT ARE THE ODDS OF BEING HIT BY A HURRICANE?

In the 30 year period from 1948 through 1977 a total of 46 hurricanes reached the United States coast. In total they killed 1,647 people. This ranged from a high of 395 in 1957 (hurricane Audrey) to a low of one fatality in 1974 (hurricane Carmen). In an area subject to hurricanes, these are the odds of one hitting in any given year.

The Odds: **228-1** against

Source: Department of Commerce, *Tropical Cyclones of the North Atlantic Ocean 1871-1977*, Asheville, North Carolina, 1978, pp. 22, 25-28.

WHAT ARE THE ODDS ON BEING HIT BY A TORNADO?

As with all weather based phenomena and earthquakes your location is key. If you live in Rhode Island or Vermont the odds are against one. Those states haven't reported a tornado in years. If you're a Texan the odds of a tornado coming your way are way up there; the average number of tornadoes each year is 150. That's 10 times the Kansas rate, the Wizard of Oz notwithstanding.

The Odds: against

	Average number of tornadoes per year 1976-1977	Of a tornado somewhere in in the state on any given day
Alabama	25.5	13.3-1
Alaska	0	—
Arizona	3.5	103.3-1
Arkansas	29.0	11.6-1
California	4.5	80.1-1
Colorado	37.0	8.9-1
Connecticut	0	—
Delaware	2.5	145.0-1
Washington, D.C.	0	—
Florida	51.0	6.1-1
Georgia	18.5	18.7-1
Hawaii	0	—
Idaho	0	—
Illinois	30.0	11.2-1
Indiana	20.5	16.8-1
Iowa	27.0	12.5-1
Kansas	15.0	23.3-1
Kentucky	8.5	41.9-1
Louisiana	28.0	12.0-1
Maine	1.0	364.0-1
Maryland	3.0	120.7-1

The Odds: against

	Average number of tornadoes per year 1976-1977	Of a tornado somewhere in in the state on any given day
Massachusetts	1.5	242.3-1
Michigan	32.0	10.4-1
Minnesota	18.0	19.3-1
Mississippi	34.5	9.6-1
Missouri	14.5	24.2-1
Montana	4.0	90.2-1
Nebraska	47.0	6.8-1
Nevada	1.0	364.0-1
New Hampshire	1.0	364.0-1
New Jersey	1.5	242.3-1
New Mexico	5.0	72.8-1
New York	6.5	55.1-1
North Carolina	25.0	13.6-1
North Dakota	41.0	7.9-1
Ohio	16.0	21.8-1
Oklahoma	41.0	7.9-1
Oregon	0	——
Pennsylvania	18.0	19.3-1
Puerto Rico	0	——
Rhode Island	0	——
South Carolina	13.0	27.1-1
South Dakota	16.0	21.6-1
Tennessee	12.0	29.4-1
Texas	150.5	1.4-1
Utah	0	——
Vermont	0	——
Virginia	10.5	33.8-1
Virgin Islands	0.5	729.0-1
Washington	0	——
West Virginia	1.5	242.3-1
Wisconsin	13.5	26.0-1
Wyoming	18.5	18.7-1

Source: Department of Commerce, National Oceanic and Atmospheric Administration, Environmental Data Service, National Climate Center, *General Summary of Tornadoes*, 1977.

WHAT ARE THE ODDS ON A STATE HAVING AN EARTHQUAKE?

This depends entirely on where you are. If your home happens to be in California's San Andreas Mountains your odds couldn't be better, since you're atop one of the world's most hazardous faults. On the other hand if you're in Florida you have little to worry about from earthquakes. They haven't recorded one since 1973 and that was minute, so you can fret about hurricanes instead.

In the last two years the National Earthquake Information Center has reported 749 earthquakes in 35 different states.

States can be enormous while earthquakes tend to be localized affairs. Clearly this affects the local odds. We took the average number of earthquakes for 1977 and 1978 on a state by state basis for the 35 states that had earthquakes to compute the odds.

The Odds: against unless noted

Alabama	11.0-1	Nebraska	23.0-1
Alaska	4.2-1 for	Nevada	3.5-1
Arizona	3.6-1	New Hampshire	23.0-1
Arkansas	11.0-1	New Jersey	23.0-1
California	9.2-1 for	New Mexico	11.0-1
Colorado	5.0-1	New York	5.0-1
Delaware	23.0-1	North Carolina	7.0-1
Georgia	23.0-1	Ohio	23.0-1
Hawaii	10.6-1 for	Oklahoma	7.0-1
Idaho	2.7-1	Oregon	23.0-1
Illinois	5.0-1	Pennsylvania	11.0-1
Maine	11.0-1	South Carolina	3.8-1
Maryland	11.0-1	Tennessee	7.0-1
Massachusetts	23.0-1	Texas	3.8-1
Mississippi	11.0-1	Utah	Even
Missouri	5.0-1	Virginia	11.0-1
Montana	Even	Wyoming	1.3-1 for

Source: Department of the Interior, *Geological Survey*, Denver, Colorado.

Chapter

IT'S CRIMINAL
Crime, Victims, Courts and Prisoners

Count: 1—2—3. In those three seconds another crime has taken place in America. As this section points out about nine out of 10 of these crimes are nonviolent. Yet on the average day we have 1,529 aggravated assaults, 1,142 robberies, 184 rapes and 54 murders. Twice a minute someone somewhere is in real physical danger, hurt or dead.

Most of the following figures are taken from the F.B.I.'s 1978 Uniform Crime Report, which profiles seven major crime categories based on statistics filed by law enforcement agencies across the nation.

This section makes for sobering reading. The annual total was over 11 million reported crimes. What's more, other Justice Department studies indicate that only about half of all crimes are reported. If so, one out of 10 Americans ends up as a crime victim each year.

Our section starts by examining where certain crimes occur. You'll see that murder is a southern specialty while the West Coast is way above average on rape. You'll also discover that some of our most popular vacation playgrounds are pretty deadly places.

Next we really do bring the figures home to you. For all seven categories of crime we calculated the odds for virtually every locale in the United States. If you happen to be one of the one out of six families who'll move this year, these figures may play a part in where you relocate.

More detailed odds on murder follow. Supporters of stronger gun control legislation will find much of interest here. Guns account for over six out of 10 deaths. Other highlights of this section include the high mortality rate among the "other man or woman" in romantic triangle murders and the exceptionally high murder rate among black males in their early 20's. It beats out all diseases and other forms of accidents as the top cause of death among this group.

From murder we turn to rape, a crime that takes place every seven and a half minutes. There are a number of fairly startling figures here. One example: in half of all completed rapes the man is known to the woman.

Robbery, recovery rates and car theft follow. It's small wonder that car insurance rates have rocketed. About a million autos are stolen each year.

Finally we look at the two million citizens who are in trouble with the law. You'll see why you're most apt to beat a murder rap and why canny defense lawyers prefer jury trials.

When you've been through the chapter we think you'll agree that the overall picture is really criminal.

WHAT ARE THE ODDS OF FINDING A PIG IN A POKE?

There are about 100 policemen in jail at the moment. According to the Bureau of Prisons there are 436,000 people in the pokey (both prison and jail).

The Odds: 4,359-1

Source: American Federation of Police, Bureau of Prisons.

WHAT ARE THE ODDS ON FLYING THE COOP?

Not all that bad if you're a prisoner in a state or federal penitentiary. In 1975 out of a total prison population of 240,393 some 8,582 prisoners escaped.

The Odds:

Escaping from a state or federal prison	**27.0-1** against	
Escaping from a state prison	**26.3-1** against	
Escaping from a federal prison	**35.3-1** against	

Source: Sheri D. Touchstone, *Comparison Analysis of Escapes from State and Federal Correctional Institutions 1972 through 1975*, Research, Planning and Development Division, Technical Note No. 55, Huntsville, Texas Department of Corrections, 1978, pp. 3, 4.

WHAT ARE THE ODDS OF BEING SOLD DOWN THE RIVER?

In 1978 6,129 people took the Delta Queen paddle steamer down the river from Cincinnati.

The Odds: **35,533-1** against

Source: Data from the Delta Queen.

WHAT ARE THE ODDS OF BEING A THREE TIME LOSER?

A special 1974 survey done among inmates of state correctional institutions showed that 19% had served three sentences, 12% four, and 16% five or more terms.

The Odds: **1.1-1** against

Source: Department of Justice, *Survey of Inmates of State Correctional Facilities 1974*, Special Report SD-NPS-SR-2, pp. 35, 36.

WHAT ARE THE ODDS ON BEING VICTIMIZED BY A MAJOR CRIME BY GEOGRAPHICAL AREA?

The United States is broken down into 255 greater metropolitan areas (known as standard metropolitan statistical areas). The F.B.I. receives and publishes reports in all of these and compares them with their population to determine actual relative crime rates unbiased by differing city sizes.

Crime rates are affected by many other factors ranging from unemployment to the character of the city and the number of visitors entertained each year.

Because of this it's perhaps not so surprising to find that Las Vegas is, in terms of crimes per resident, the most dangerous city in America today. You could justifiably feel you're gambling on trouble since the odds there are only 9.6-1 against being involved in one of the F.B.I.'s major crime categories each year.

By way of contrast our safest city is Johnstown, Pennsylvania which is six times safer than Las Vegas and three times safer than the average United States city.

When the two cities are compared some startling contrasts emerge. For example you're 12 times as apt to be raped in Las Vegas as in Johnstown. Given that half the population is male the true picture in Nevada's famed city is one rape for every 738 resident women. By the same token you're 11 times more likely to be robbed, and that doesn't factor in the one-armed bandits.

The safest and most dangerous cities in America are as follows:

The Odds: against

Safest areas		Most dangerous areas	
Johnstown, PA	58.0-1	Las Vegas, NV	9.6-1
Steubenville-Weirton OH/WV	45.0-1	Phoenix, AZ	10.5-1

Safest areas		Most dangerous areas	
Johnson City-Kingsport Bristol TN/VA	44.1-1	Daytona, Beach, FL	10.6-1
St. Cloud, MN	42.3-1	Orlando, FL	10.9-1
Eau Claire, WI	42.0-1	Miami, FL	10.9-1
Wheeling, WV	41.5-1	Tucson, AZ	11.0-1
Lancaster, PA	38.7-1	Sacramento, CA	11.0-1
North East, PA	36.2-1	San Fran.-Oakland, CA	11.4-1
Altoona, PA	36.0-1	Bakersfield, CA	11.6-1
Utica-Rome, NY	35.6-1	Fresno, CA	11.7-1

When you consider that six of the 10 least safe cities are major tourist resorts you may be tempted to wonder if some vacations aren't more exciting than was planned.

Source: Department of Justice, *F.B.I. Uniform Crime Report.*

WHAT ARE THE ODDS ON VARIOUS CRIMES BEING REPORTED TO THE POLICE?

For every 100 rapes 47 aren't even reported to the police. Six out of 10 burglaries go unreported, seven out of 10 pickpockets or purse snatchers are also simply not recorded because the victims don't report them. It's a sad comment on the lethargy of us all that just over half of all violent crime (51.2%) is never reported to the police.

The Odds: for unless noted

All crimes of violence	Even
Rape	1.1-1
Robbery	1.1-1
Aggravated assault	1.4-1
All crimes of theft	2.8-1 against
Burglary	1.1-1 against
Motor vehicle theft	2.3-1

Source: Ibid.

WHAT ARE THE ODDS ON YOU, YOUR HOME, YOUR PLACE OF WORK OR YOUR CHILD'S SCHOOL BEING BOMBED?

Every 10 hours a bomb goes off somewhere in America. These added up to total of 867 bombings in 1977. For every two bombs that go off another one is attempted which fails. If this sounds like a lot of mayhem it is: there were 1,308 bombings in 1977. Yet bombing is on the decline. 1977 had only two thirds of 1975's rate of 2,074. The odds of being bombed vary widely by state. Based on previous incidents you're in good shape in Vermont or North Dakota—no bombings in 1977 at all (while South Dakota had four successful attempts). On the other hand you wouldn't want to be in California—221 actual or attempted bombings— New York with 130 or Illinois with 121. These three states alone accounted for just over one third of all bombing incidents, and very nearly four out of 10 actual bombings (38%).

The disparity between attempted bombings and success is interesting. Bombers in Illinois are remarkably successful at actually pulling off the blast (93% succeed) while Californians fail with regularity (44% fail). Can it be that loud explosions inhibit a laid back lifestyle? New York State falls somewhere in between; one out of four bombs proves to be a dud.

When people set out to bomb something or someone, what are they after? Residences account for three out of 10 actual bombings, business premises another 26%, schools a worrisome 10% and individuals 6% (these almost always occur in their cars). These four categories add up to over seven out of 10 bombings. Other specific targets are government property, notably local government buildings, banks, public utilities and the post office.

The Odds: against

Being personally bombed	3,887,000-1
Having your house bombed	292,885-1

Source: Department of Justice, F.B.I. Uniform Crime Report, *Bomb Summary*, 1977.

WHAT ARE THE ODDS OF BEING MURDERED IN DIFFERENT PARTS OF THE COUNTRY?

There were 19,560 murders in the United States in 1978. An amazing 40% of all murders took place in the southern states, which account for just 32% of our population. Murder is a seasonal pasttime. December is the big month (more than 15% above average), while the first six months of each year (notably February and June) have fewer than average homicides.

The Odds: against

Each year

Nationally	11,111-1	North central	14,083-1
Northeast	14,492-1	South	8,620-1
		West	10,525-1

Source: Department of Justice, *F.B.I. Uniform Crime Report*.

WHAT ARE THE ODDS ON BEING KILLED IN THE CITY OR THE COUNTRY?

Murders occur least frequently in small towns, 5.2 murders per 100,000 inhabitants, and most frequently in big cities, 9.9 murders per 100,000 inhabitants.

The Odds: of being murdered each year all against

Metropolitan areas	10,100-1
Small cities outside metropolitan areas	19,230-1
Rural areas	13,332-1

Source: Ibid.

WHAT ARE THE ODDS, BY REGION, ON HOW A MURDER IS COMMITTED?

The weapon in almost one out of two murders is a handgun, while over all some form of firearm accounts for 63% of all murders. Parts of the assailant's body such as fists or feet accounted for six out of 100 murders, a total of almost 1,200 murders in the year under discussion. Favorite weapons vary widely by regions. In the South you're most apt to be shot; in the Northeast you're most likely to be stabbed.

The Odds: against unless noted

	Firearms	Cutting or Stabbing	Personal Weapon Hand, feet, etc.	Other Weapon Club, poison, etc.
Northeast	1.1-1 for	3.0-1	9.6-1	6.3-1
North Central	2.0-1 for	5.2-1	17.2-1	7.8-1
South	2.3-1 for	5.4-1	21.2-1	8.7-1
West	1.3-1 for	3.5-1	15.4-1	5.9-1

Source: Ibid.

WHAT ARE THE ODDS ON THE MURDER VICTIM'S RELATIONSHIP TO THE MURDERER IN ROMANTIC-TRIANGLE MURDERS?

In three-way affairs it's the third party that ends up in the morgue. A wife is more apt to do in her husband or his acknowledged girlfriend than the other way around.

The Odds: against unless noted

The husband	22.8-1
The wife	28.4-1
An immediate family member	51.6-1
The third party (acquaintance)	1.1-1 for
Acknowledged boyfriend	39.0-1
Acknowledged girlfriend	26.7-1

Source: Ibid.

WHAT ARE THE ODDS, BASED ON SEX AND RACE, OF BEING SLAIN?

In three out of four cases the murder victim is a man. Out of every 100 murders 49 victims are white, 48 are black and three are of other races. So blacks, who comprise about 11.5% of the population, account for over four times their population proportion in numbers of murder victims.

The Odds: against

Men	7,452-1
Women	25,102-1
Blacks	2,999-1
Whites	18,326-1

Source: Ibid.

WHAT ARE THE ODDS ON THE OFFENDER BEING THE SAME SEX AND RACE AS THE MURDER VICTIM?

As shown by the figures below, the vast majority of murders are committed by men against members of their own race.

The Odds: against unless noted

Victim	Offender			
	White man	White woman	Black man	Black woman
White man	1.8-1 for	12.3-1	3.2-1	75.9-1
White woman	1.2-1 for	8.3-1	2.4-1	40.7-1
Black man	9.5-1	165.7-1	4.0-1 for	12.5-1
Black woman	26.0-1	None	3.0-1 for	3.6-1

Source: Ibid.

WHAT ARE THE ODDS THAT A MURDER WILL BE UNDERTAKEN BY ONE VERSUS MORE THAN ONE ASSAILANT?

Almost nine out of 10 murders (89%) involve a single assailant.

The Odds: **8.1-1** for

Source: Ibid.

WHAT ARE THE ODDS ON A MURDERER BEING A CERTAIN AGE?

Well over 50% of all arrested murderers are under 30 years old and over one in 100 is under the age of 15.

The Odds: against unless noted

Under 15	75.8-1
15-29 Cumulative	1.5-1 for
15-19	4.6-1
20-24	3.1-1
25-29	4.4-1
30-34	7.4-1
35-39	10.9-1
40-44	16.1-1
45-49	22.9-1
50-59	18.5-1
60 +	35.8-1

Source: Ibid.

WHAT ARE THE ODDS ON A MURDER VICTIM BEING A CERTAIN AGE?

While murder victims are also concentrated in the 15 to 30 age group a depressingly large number are under 15 (5%) and over 60 (5.9%).

The Odds: against

Under 1	89.9-1	30-34	7.5-1
1-9	37.5-1	35-39	10.0-1
10-14	75.9-1	40-44	12.5-1
15-19	10.5-1	45-49	16.2-1
20-24	5.1-1	50-59	9.6-1
25-29	5.2-1	60 +	15.9-1

Source: Ibid.

WHAT ARE THE ODDS ON THE REASONS FOR A MURDER?

Almost half of all murders (45.5%) occur as the direct result of arguments, notably arguments between family members and friends.

The Odds: against

Robbery	8.8-1
Narcotics felony	57.8-1
Sex offense	70.4-1
Other felony	29.3-1
Suspected felony	16.8-1
Arguments—romantic triangles	36.0-1
Influence of alcohol/narcotics	17.9-1
Property or money	27.6-1
Other arguments	1.9-1
Other motives & circumstances	4.5-1
Unknown	6.2-1

Source: Ibid.

WHAT ARE THE ODDS ON A MURDER BEING SOLVED?

In terms of an arrest being made, excellent; 76% of all murders are cleared by law enforcement.

The Odds: **3.2-1 for**

Source: Ibid.

WHAT ARE THE ODDS THAT A RAPE WILL BE REPORTED?

Every eight minutes another forcible rape is reported in the United States. That's an average of 180 every day, 67,131 for all of 1978.

The incidence of rape has grown upsettingly in the past 10 years. Today 60 out of every 100,000 women are forcibly raped each year, a 66% increase over 1969 figures. What's even more saddening is that these are only reported figures. Most authorities are convinced that the actual rape figures are at least twice as high as those reported. That would take the number to a conservative 120 victims per 100,000. Even this figure doesn't reflect the full measure of violations done to women because forcible incest is not included.

The Odds: nationally

833.3-1 against estimated total
1,666.0-1 against reported total

Source: Ibid.

WHAT ARE THE ODDS ON BEING RAPED IN THE CITY OR THE COUNTRY?

Rape occurs most frequently in greater metropolitan areas where there are 72 victims per 100,000 females. The comparable figure for smaller towns is 31, for rural areas 27.

The Odds: of a reported rape each year (against)

Metropolitan areas	1,388-1
Smaller cities outside Metropolitan areas	3,255-1
Rural areas	3,702-1

Source: Ibid.

WHAT ARE THE ODDS ON WHERE A RAPE WILL OCCUR?

The West is this nation's black spot for rape. With less than 20% of our population these states account for over one out of four reported rapes (27.2% of the total).

Alaska is the worst of these states while Nevada, California and Colorado all have bad records too.

Here are the odds on a regional basis as well as for certain chosen high offense states and the two states with the best records, North Dakota and New Hampshire.

The Odds: of a woman being forcibly raped (against)

Region	
New England	2,231-1
North Central	2,011-1
South	1,591-1
West	1,126-1
Selected states	
Michigan	1,294-1
Florida	1,111-1
Colorado	1,033-1
Nevada	950-1
Alaska	921-1
California	1,008-1
North Dakota	5,746-1
New Hampshire	5,462-1

Source: Ibid.

WHAT ARE THE ODDS ON A RAPE VICTIM BEING WITHIN CHOSEN AGES?

Out of every 1,000 women about three are raped each year. As the figures show the early twenties are the most dangerous years for women.

The Odds:

Age of victim

12-15	908-1 against	20-24	384-1 against
16-19	475-1 against	25-34	832-1 against

Source: Department of Justice, National Criminal Justice and Information Service, Law Enforcement Assistance Administration, *Criminal Victimization in the United States*, 1976, p. 25.

WHAT ARE THE ODDS ON THE RACE AND MARITAL STATUS OF RAPE VICTIMS?

While roughly one out of nine females are black they suffer one out of four rapes. Rape is far more prevalent among divorced and single women. This is rarely because they act in a "provocative manner." Rather it's because for both social and work reasons they tend to be out more at night (when rapists prey) and because they're more apt to be alone.

The Odds: against

Race

White	1,428-1
Black	525-1

Marital Status

Never married	587-1
Married	2,499-1
Divorced	454-1

Source: Ibid., pp. 26, 29.

WHAT ARE THE ODDS ON A WOMAN BEING RAPED BY SOMEONE SHE KNOWS, COMPARED TO STRANGERS, UNDER DIFFERENT CONDITIONS?

Women are twice as apt to confront a stranger in an attempted rape as someone they know at least by sight. That's attempted; one out of two completed rapes is done by a non-stranger. Also, half of all rapes where the attacker is known take place inside the victim's home.

The Odds:

Attempted rape involving a stranger	2,499-1
Attempted rape involving non-strangers	4,999-1
Rape involving strangers	9,999-1
Rape involving non-strangers	9,999-1

Of those committed:

In own house – involving strangers	50-1
In own house – involving non-strangers	Even

Source: Ibid., pp. 43, 56.

WHAT ARE THE ODDS ON WHERE A RAPE WILL OCCUR?

One out of three rapes occurs out of doors on a street, park, playground/schoolground or parking lot, while over one out of four occurs inside a woman's home. In combination over six out of 10 rapes take place in these two locales.

The Odds: against

Inside victim's home	2.7-1
Near victim's home	7.1-1
Inside non-residential building	15.1-1
Inside school	36.3-1
On street/park/playground/ schoolground/parking lot	2.0-1
Elsewhere	4.7-1

Source: Ibid., p. 55.

WHAT ARE THE ODDS ON THE RACE OF A RAPIST?

There were 28,155 arrests for forcible rape in 1978, a 42% arrest to reported crime ratio. Thus, assuming that again as many forcible rapes go unreported, about three out of four such crimes go unpunished. Of those arrested 48% were white, 48% black and 2% from other races. The black arrest rate is sobering when one considers that blacks account for less than 12% of the population.

The Odds:

White	**1.1-1** against
Black	**1.1-1** against
Other	**49.0-1** against

WHAT ARE THE ODDS THAT AN ARRESTED RAPIST WILL BE IN A CERTAIN AGE BRACKET?

Well over half of all rapes are committed by men under the age of 25. Boys under the age of 15 commit more rapes than men over the age of 50. Last year 75 boys under the age of 10 were arrested for rape.

The Odds: against unless noted

Under 15	24.6-1
15-19	3.3-1
20-24	2.7-1
25-29	4.3-1
15-29	Even
30-34	7.8-1
35-39	13.9-1
40-44	26.0-1
45-49	40.7-1
50 +	32.3-1

Source: Ibid.

WHAT ARE THE ODDS, BY AREA OF THE COUNTRY, ON BEING ROBBED IN A YEAR?

In 1978 there were 417,038 robberies in the United States. That's one every 76 seconds and amounts to 191 per 100,000 people. Robbery is most prevalent in the East (31% of total robberies and 22% of the population), least prevalent in the West. This is probably in part explained by the high percentage of robberies which are street crimes: almost 50%. Out in the wide open spaces you don't have lonely city streets and pedestrians, the robber's ideal environment.

The Odds: against

Nationally	523-1
New England	376-1
North Central	657-1
South	661-1
West	438-1

Source: Department of Justice, *F.B.I. Uniform Crime Report,* 1978.

WHAT ARE THE ODDS ON BEING ROBBED IN THE CITY OR THE COUNTRY?

Seven out of 10 robberies occur in cities with a population of over 100,000.

The Odds:		
Cities of 100,000 +	**200-1** against	
Other cities	**1,999-1** against	
Rural areas	**4,761-1** against	

Source: Ibid.

WHAT ARE THE ODDS ON THE TYPE OF WEAPON A ROBBER USES?

The strong-arm tactics of a mugger are about as prevalent as firearms (37% to 41%) in this crime.

The Odds: Weapon in a robbery

Firearm	1.4-1 against
Strong-arm	1.7-1 against
Knife	6.7-1 against
Other weapon	10.1-1 against

Source: Ibid.

WHAT ARE THE ODDS ON THE LOCALE OF A ROBBERY?

While street crimes are the largest group of robberies a number of other types are also common.

The Odds: against

Street & highway	1.1-1
Commercial	5.9-1
Convenient store	13.3-1
Miscellaneous	6.5-1
Home	7.8-1
Bank	82.3-1
Gas station	16.8-1

Source: Ibid.

WHAT ARE THE ODDS ON YOUR HOME BEING ROBBED?

That household figure may look small out of context, but 47,125 homes were invaded by robbers in 1978. There were about 74,500,000 households in mid-1978.

The Odds: **1,580-1** against

Source: Ibid.

WHAT ARE THE ODDS ON A BANK OR GAS STATION BEING HELD UP?

Banking is normally regarded as a staid business but it has its exciting moments. In 1978 there were 5,004 bank stick-ups compared to 52,604 banks. During the same period there were 23,352 gas station heists among the country's 177,800 outlets. As many rueful managers can attest, some stations were hit more than once, so the odds are somewhat longer than we show.

The Odds:

Bank	9.5-1 against
Gas station	6.6-1 against

Source: Ibid..

WHAT ARE THE ODDS ON STOLEN MONEY BEING RECOVERED?

In 1978 all robberies accounted for losses of $181 million with an average loss of $434. Less than $47 million of this was recovered, since only 26% of these crimes resulted in an arrest (which doesn't necessarily mean recovery of the money or valuables).

The Odds: **2.8-1** against

Source: Ibid.

WHAT ARE THE ODDS OF AN ARRESTED ROBBERY SUSPECT HAVING CERTAIN CHARACTERISTICS?

Arrested robbers are usually male (93%) and tend to be black (59%). One out of six are 15 or under; half of those are in their teens and almost 90% are under the age of 30.

The Odds:

Male	13.3-1 for
White	1.6-1 against
Black	1.4-1 for
Other race	26.8-1 against

Age	Cumulative odds
15 or under	5.2-1 against
16-19	Even
20-21	1.6-1 for
22-24	2.9-1 for
25-29	6.9-1 for

Source: Ibid.

IF YOU ARE ROBBED WHAT ARE THE ODDS ON THE MONETARY VALUE OF WHAT'S TAKEN?

In about half of all robberies $50 or less is taken.

The Odds: against

No monetary value	44.5-1
Less than $10	4.6-1
$10-$49	2.5-1
$50-$99	5.5-1
$100-$249	5.4-1
$250+	6.7-1

Source: Ibid.

IF YOU ARE ROBBED WHAT ARE THE ODDS ON RECOUPING SOME OR ALL OF YOUR LOSS?

In over two out of three robberies none of the loss is ever recovered while in just under one in five full recovery is made.

The Odds: against

Recovering anything	2.2-1
Some recovered	7.3-1
Less than half recovered	29.3-1
Half or more recovered	10.6-1
All recovered	4.2-1

Source: Ibid.

WHAT ARE THE ODDS, BY RACE, ON BEING BURGLARIZED OR HAVING YOUR MOTOR VEHICLE STOLEN?

Blacks suffer far more from these crimes than whites or those of another ethnic background. Presumably in part the latter have the lowest rate because they often live in tightly-knit communities.

The Odds: against

	Burglary	Motor vehicle stolen	
All races	10.2-1	59.6-1	
White	10.9-1	61.9-1	
Black	6.6-1	45.5-1	
Other	12.9-1	69.4-1	

Source: Ibid.

WHAT ARE THE ODDS ON YOUR HOME BEING BURGLARIZED WHEN YOU RENT VERSUS OWN YOUR DWELLING?

Very nearly 50% more burglaries occur in rented houses than in owned ones (117 per 1,000 rented households versus 73.3 owned ones).

The Odds:	Rented	**7.5-1** against
	Owned	**12.6-1** against

Source: Ibid.

WHAT ARE THE ODDS ON A COMMERCIAL ESTABLISHMENT BEING BURGLARIZED BASED ON THE TYPE OF BUSINESS IT IS AND THE NUMBER OF EMPLOYEES IT HAS?

Out of every 1,000 establishments a staggering 217 are burglarized in a given year. The rate is higher for wholesale establishments and for mid-size firms.

The Odds: against

All establishments	3.6-1
Retail establishments	2.5-1
Wholesale establishments	2.2-1
Service establishments	4.6-1
Other establishments	5.4-1

Average number of paid employees

1-3	4.1-1
4-7	3.0-1
8-19	2.6-1
20 +	2.7-1

Source: Ibid.

WHAT ARE THE ODDS ON BEING IN TROUBLE WITH THE LAW?

In 1976 just under two million Americans, including half a million juveniles, were under some kind of formal supervision. Of this total 457,528 were in some form of prison. We computed these odds based on all those aged 10 or over during this period.

The Odds: against

On probation, parole or confined	92.1-1
On probation	142.0-1
On parole	852.0-1
Confined	390.0-1

Source: Department of Justice, Law Enforcement Assistance Administration, *State and Local Probation and Parole Systems*, No. (SD) P-1.

WHAT ARE THE ODDS BY TYPE OF CRIME OF AN ACCUSED CRIMINAL BEING CONVICTED AND SENTENCED?

An analysis of all criminal cases tried in U.S. District Courts for the 12 month period ending June 30, 1978 was used to derive the figures. These clearly show that, at least in Federal Courts, accused murderers have the greatest chance of beating the rap, while accused robbers are generally convicted.

The Odds: for

All criminal offenses	3.9-1
1st degree murder	1.7-1
Rape	3.1-1
Drunk driving and traffic offenses	3.2-1
Burglary	3.4-1
Auto thefts	4.8-1
Robbery	6.6-1

Source: Department of Justice, Administrative Office of the United States Courts, *1978 Annual Report of the Director*, pp. A-73, A-74, A-75.

WHAT ARE THE ODDS OF A CONVICTION ON VARIOUS TYPES OF FEDERAL CHARGES BEING OVERTURNED ON APPEAL?

We analyzed all cases reversed or denied in 1977 and 1978 in the U.S. Federal Courts. Just under one in five convictions were thrown out of court on appeal.

The Odds: against

All appeals	5.3-1
Civil appeals	4.5-1
Criminal appeals	8.5-1
Bankruptcy appeals	5.5-1
Administration appeals	3.9-1

Source: Department of Justice, Administrative Office of the United States Courts, *1978 Annual Report of the Director.*

WHAT ARE THE ODDS ON A DEFENDANT BEING ACQUITTED IN A HOMICIDE CASE WHICH IS TRIED BY A JURY?

Surprisingly the odds of beating a murder rap are better if you're up for premeditated murder (first degree) than unpremeditated. Only one out of 10 second degree murderers escaped conviction, while 42% of those up for manslaughter were found not guilty.

The Odds:

All homocides	2.2-1 against
1st degree murder	1.9-1 against
2nd degree murder	9.9-1 against
Manslaughter	1.3-1 against

Source: Ibid.

WHAT ARE THE ODDS ON A CRIMINAL DEFENDANT GETTING OFF WITH A JURY TRIAL VERSUS A COURT TRIAL?

Of the 7,016 criminal cases which came to trial during our analysis period fully 75% were jury trials. The odds below clearly show why defense attorneys favor juries.

The Odds:	Jury trial	**2.9-1** against
	Court trial	**4.6-1** against

Source: Ibid.

WHAT ARE THE ODDS OF THE MOST FREQUENT CRIMES GEOGRAPHICALLY?

We'll leave it up to you where you want to live.

Standard Metropolitan Statistical Area	Crime Index total	Murder and non-negligent man-slaughter	Forcible rape	Robbery
Abilene, TX .... Includes Callahan, Jones and Taylor Counties	24.2	14924	2993	1530
Akron, OH .... Includes Portage and Summit Counties	19.4	23808	2746	857
Albany, GA .... Includes Dougherty and Lee Counties	18.6	8063	2226	706
Albany—Schenectady— Troy, NY .... Includes Albany, Montgomery, Rensselaer, Saratoga and Schenectady Counties	25.9	55534	7691	1561
Albuquerque, NM .... Includes Bernalillo and Sandoval Counties	14.5	8402	1662	497
Alexandria, LA .... Includes Grant and Rapides Parishes.	24.1	6368	5616	1505
Allentown—Bethlehem— Easton, PA—NJ .... Includes Carbon, Lehigh and Northampton Counties, PA, and Warren County, NJ	32.0	38460	9432	1578
Altoona, PA .... Includes Blair County	36.0	3332	6288	1830
Amarillo, TX .... Includes Potter and Randall Counties	16.7	12194	2687	991
Anaheim—Santa Ana— Garden Grove, CA .... Includes Orange County	14.6	23809	2832	576
Ann Arbor, MI .... Includes Washtenaw County	15.7	12820	2550	859
Anniston, AL .... Includes Calhoun County	26.8	8064	4524	1452
Appleton—Oshkosh, WI ...... Includes Calumet, Outagamie and Winnebago Counties	23.4	71428	12820	5987
Asheville, NC .... Includes Buncombe and Madison Counties	31.9	9433	6288	1649

Standard Metropolitan Statistical Area	Aggra-vated assault	Burglary	Larceny-theft	Motor vehicle theft
Abilene, TX	768	114.4	37.8	483
Akron, OH	407	94.6	31.5	268
Albany, GA	333	55.6	33.9	457
Albany—Schenectady—Troy, NY	558	89.8	43.2	457
Albuquerque, NM	221	51.1	25.3	253
Alexandria, LA	347	95.9	38.4	577
Allentown—Bethlehem—Easton, PA—NJ	1439	116.8	51.0	506
Altoona, PA	976	102.3	71.3	565
Amarillo, TX	306	61.0	27.3	282
Anaheim—Santa Ana—Garden Grove, CA	496	43.8	29.3	206
Ann Arbor, MI	365	72.6	25.9	201
Anniston, AL	372	89.1	53.5	358
Appleton—Oshkosh, WI	2550	115.0	31.9	709
Asheville, NC..............	472	111.0	63.2	347

*All odds are presented excluding −1.

Standard Metropolitan Statistical Area	Crime Index total	Murder and non-negligent man-slaughter	Forcible rape	Robbery
Atlanta, GA .... Includes Butts, Cherokee, Clayton, Cobb, De Kalb, Douglas, Fayette, Forsythe, Fulton, Gwinnett, Henry, Newton, Paulding, Rockdale, and Walton Counties	13.4	7193	1720	320
Atlantic City, NJ .... Includes Atlantic County	13.0	11764	2557	422
Augusta, GA—SC .... Includes Columbia and Richmond Counties, GA and Aiken County, SC	17.7	6328	1926	599
Austin, TX .... Includes Hays, Travis and Williamson Counties	14.1	9433	1956	671
Bakersfield, CA .... Includes Kern County	11.6	5951	2231	449
Baltimore, MD .... Includes Baltimore City and Anne Arundel, Baltimore, Carroll, Harford and Howard Counties	14.2	8546	2450	223
Baton Rouge, LA .... Includes Ascension, East Baton Rouge, Livingston and West Baton Rouge Parishes	12.6	8928	2293	834
Battle Creek, MI .... Includes Barry and Calhoun Counties	19.7	12194	3343	1148
Bay City, MI .... Includes Bay County	20.0	10203	3845	1350
Beaumont—Port Arthur— Orange, TX .... Includes Hardin, Jefferson and Orange Counties	17.6	6756	3278	643
Billings, MT .... Includes Yellowstone County	18.5	34482	4273	1483
Biloxi—Gulfport, MS .... Includes Hancock, Harrison and Stone Counties	26.3	10308	3194	825
Binghamton, NY—PA .... Includes Broome and Tioga Counties, NY and Susquehanna County, PA	29.7	24999	13157	4236

Standard Metropolitan Statistical Area	Aggra-vated assault	Burglary	Larceny-theft	Motor vehicle theft
Atlanta, GA	296	47.1	27.3	172
Atlantic City, NJ	364	43.1	25.8	164
Augusta, GA—SC	314	53.7	37.1	307
Austin, TX	619	50.7	24.1	293
Bakersfield, CA	230	41.1	22.3	169
Baltimore, MD	190	63.8	27.5	201
Baton Rouge, LA	184	51.0	22.4	227
Battle Creek, MI	416	72.0	33.3	529
Bay City, MI	300	93.2	31.7	531
Beaumont—Port Arthur—Orange, TX	296	64.4	32.6	314
Billings, MT	475	85.8	29.4	265
Biloxi—Gulfport, MS	209	63.2	45.6	281
Binghamton, NY—PA	1343	107.0	47.3	670

*All odds are presented excluding −1.

Standard Metropolitan Statistical Area	Crime Index total	Murder and non-negligent man-slaughter	Forcible rape	Robbery
Birmingham, AL .... Includes Jefferson, St. Clair, Shelby and Walker Counties	17.2	6578	2840	605
Bloomington, IN .... Includes Monroe County.	18.8	23255	4183	2557
Boise, ID .... Includes Ada County	15.0	47618	2544	1207
Boston, MA .... Includes Essex, Middlesex, Norfolk and Suffolk Counties	17.7	21738	3801	413
Bradenton, FL .... Includes Manatee County	18.3	22221	4443	1221
Bridgeport, CT .... Includes Fairfield County	19.0	18867	5746	722
Brockton, MA .... Includes Plymouth County	17.2	99999	5347	1505
Brownsville—Harlingen— San Benito, TX .... Includes Cameron County	19.1	12047	6060	2040
Bryan—College Station, TX.... Includes Brazos County	17.9	8620	2975	1461
Buffalo, NY .... Includes Erie and Niagara Counties	20.0	21276	4183	572
Burlington, NC .... Includes Alamance County	32.1	14084	6210	3424
Canton, OH .... Includes Carroll and Stark Counties	26.0	30302	3558	852
Cedar Rapids, IA .... Includes Linn County	14.6	41666	6666	1703
Charleston— North Charleston, SC .... Includes Berkeley, Charleston, and Dorchester Counties	13.7	10100	2140	458
Charleston, WV .... Includes Kanawha and Putnam Counties	24.2	12986	4064	814
Charlotte—Gastonia, NC ...... Includes Gaston, Mecklenburg and Union Counties	15.6	7812	3545	716

Standard Metropolitan Statistical Area	Aggravated assault	Burglary	Larceny-theft	Motor vehicle theft
Birmingham, AL	255	65.9	34.98	162
Bloomington, IN	597	77.9	29.5	363
Boise, ID	406	53.3	26.3	259
Boston, MA	341	66.9	44.0	76.0
Bradenton, FL	328	57.0	31.8	384
Bridgeport, CT	738	72.7	37.3	136
Brockton, MA	344	56.8	38.9	115
Brownsville—Harlingen— San Benito, TX	344	72	33.5	237
Bryan—College Station, TX	359	60.7	28.0	943
Buffalo, NY	470	73.7	40.0	177
Burlington, NC	474	118.0	55.1	750
Canton, OH	571	97.3	45.2	302
Cedar Rapids, IA	612	69.5	21.8	264
Charleston— North Charleston, SC	162	49.0	28.0	229
Charleston, WV	1058	96.8	41.2	302
Charlotte—Gastonia, NC	203	53.7	30.4	303

*All odds are presented excluding − 1.

Standard Metropolitan Statistical Area	Crime Index total	Murder and non-negligent man-slaughter	Forcible rape	Robbery
Chattanooga, TN—GA....... Includes Hamilton, Marion, and Sequatchie Counties, TN and Catoosa, Dade and Walker Counties, GA	22.0	9708	6578	870
Chicago, IL .... Includes Cook, Du Page, Kane, Lake, McHenry and Will Counties	17.0	7691	3662	363
Cincinatti, OH—KY—IN ...... Includes Clermont, Hamilton and Warren Counties, OH and Boone, Campbell and Kenton Counties, KY and Dearborn County, IN	19.0	12986	3154	703
Clarksville—Hopkinsville, TN—KY...................... Includes Christian County, KY and Montgomery County, TN	23.8	11110	3447	877
Cleveland, OH .... Includes Cuyahoga, Geauga, Lake and Medina Counties	19.2	7575	2984	260
Colorado Springs, CO .... Includes El Paso and Teller Counties	15.7	22726	1313	763
Columbia, MO .... Includes Boone County	14.8	21738	2282	1008
Columbia, SC .... Includes Lexington and Richland Counties	13.7	8771	1814	538
Columbus, GA—AL .... Includes Chattahoochee County and Columbus Consolidated Government, GA and Russell County, AL	23.3	4784	4586	629
Columbus, OH .... Includes Delaware, Fairfield, Franklin, Madison and Pickaway Counties	15.9	12986	2754	478
Corpus Christi, TX .... Includes Nueces and San Patricio Counties	14.2	4999	2363	628
Dallas—Fort Worth, TX ...... Includes Collin, Dallas, Denton, Ellis, Hood, Johnson, Kaufman, Parker, Rockwall, Tarrant and Wise Counties	12.8	6535	1952	451

Standard Metropolitan Statistical Area	Aggra- vated assault	Burglary	Larceny– theft	Motor vehicle theft
Chattanooga, TN—GA.......	394	95.2	38.4	262
Chicago, IL	421	85.1	30.7	144
Cincinatti, OH—KY—IN	431	76.8	32.7	298
Clarksville—Hopkinsville, TN— KY......................	581	72.8	47.2	358
Cleveland, OH	443	80.4	46.0	105
Colorado Springs, CO	400	55.1	28.5	294
Columbia, MO	463	65.8	22.8	399
Columbia, SC..............	158	42.5	26.7	211
Columbus, GA—AL	578	70.3	49.8	268
Columbus, OH	689	58.9	28.2	231
Corpus Christi, TX...........	300	47.2	27.6	231
Dallas—Fort Worth, TX	313	48.5	23.4	204

*All odds are presented excluding − 1.

Standard Metropolitan Statistical Area	Crime Index total	Murder and non-negligent man-slaughter	Forcible rape	Robbery
Dayton, OH .... Includes Greene, Miami, Montgomery and Preble Counties	16.0	11110	3134	341
Daytona Beach, FL .... Includes Volusia County	10.6	10988	1729	421
Decatur, IL .... Includes Macon County	20.7	18181	5746	985
Denver—Boulder, CO .... Includes Adams, Arapahoe, Boulder, Denver, Douglas, Gilpin, and Jefferson Counties	12.4	10525	1691	426
Des Moines, IA .... Includes Polk and Warren Counties	15.0	10988	3875	690
Detroit, MI................. Includes Lapeer, Livingston, Macomb, Oakland, St. Clair and Wayne Counties	15.7	6578	2069	270
Dubuque, IA .... Includes Dubuque County	22.3		30302	2403
Duluth—Superior, MN—WI ... Includes St. Louis County, MN and Douglas County, WI	21.9	66666	6493	1956
Eau Claire, WI.............. Includes Chippewa County and Eau Claire County	42.0	41666	11363	8402
Elmira, NY................. Includes Chemung County	21.2	49999	14084	3635
El Paso, TX .... Includes El Paso County	17.3	17856	3459	550
Erie, PA .... Includes El Paso County	25.6	29411	3459	916
Eugene—Springfield, OR...... Includes Lane County	15.3	23809	2057	985
Evansville, IN—KY .... Includes Gibson, Posey, Vanderburgh and Warrick Counties, IN and Henderson County, KY	21.5	29411	4565	725
Fall River, MA .... Includes Bristol County	20.3	41666	10752	1024

Standard Metropolitan Statistical Area	Aggra- vated assault	Burglary	Larceny- theft	Motor vehicle theft
Dayton, OH	444	56.0	26.7	318
Daytona Beach, FL	235	35.0	17.8	196
Decatur, IL	637	76.6	31.5	543
Denver—Boulder, CO	342	42.9	21.6	152
Des Moines, IA	408	85.3	20.5	233
Detroit, MI.................	286	62.7	31.0	119
Dubuque, IA	1723	126.0	29.2	359
Duluth—Superior, MN—WI ...	1496	78.5	34.9	281
Eau Claire, WI..............	5235	194.0	58.4	697
Elmira, NY.................	680	76.5	31.8	608
El Paso, TX	690	63.4	30.0	168
Erie, PA	738	89.6	44.5	316
Eugene—Springfield, OR......	606	56.2	23.7	263
Evansville, IN—KY	407	84.1	35.4	305
Fall River, MA	539	70.7	41.2	120

*All odds are presented excluding − 1.

Standard Metropolitan Statistical Area	Crime Index total	Murder and non-negligent man-slaughter	Forcible rape	Robbery
Fargo—Moorhead, ND—MN .. Includes Cass County, ND and Clay County, MN	28.0	62499	8546	3057
Fayetteville, NC............. Includes Cumberland County	15.6	7298	2357	591
Fayetteville—Springdale, AR .. Includes Benton and Washington Counties	33.6	26315	10525	4385
Flint, MI.................. Includes Genesee and Shawassee Counties	14.9	9999	2610	554
Florence, AZ .... Includes Colbert and Lauderdale Counties	36.3	9008	9708	2036
Fort Collins, CO .... Includes Larimer County	17.9	16128	3772	3377
Fort Lauderdale— Hollywood, FL .... Includes Broward County	12.8	9258	2308	462
Fort Myers—Cape Coral, FL ... Includes Lee County	21.6	15872	2091	1098
Fort Smith, AR—OK .... Includes Crawford and Sebastian Counties, AR and Le Flore and Sequoyah Counties, OK	30.0	9803	8928	2524
Fort Wayne, IN .... Includes Adams, Allen, De Kalb and Wells Counties	22.2	21738	4366	1589
Fresno, CA................ Includes Fresno County	11.7	6535	2113	310
Gadsden, AL .... Includes Etowah County	22.0	5746	4254	880
Gainesville, FL............. Includes Alachua County	12.1	12986	1607	540
Galveston—Texas City, TX .... Includes Galveston County	16.8	7406	2187	665
Gary—Hammond— East Chicago, IN .... Includes Lake and Porter Counties	14.5	7691	2352	399

Standard Metropolitan Statistical Area	Aggra- vated assault	Burglary	Larceny- theft	Motor vehicle theft
Fargo—Moorhead, ND—MN ..	1246	160.2	37.6	433
Fayetteville, NC.............	177	43.1	33.7	235
Fayetteville—Springdale, AR ..	789	130.1	52.3	582
Flint, MI..................	174	60.2	25.1	257
Florence, AL	1148	120.0	64.9	387
Fort Collins, CO	487	78.9	25.0	443
Fort Lauderdale— Hollywood, FL	289	47.1	20.2	192
Fort Myers—Cape Coral, FL ...	250	76.5	37.5	498
Fort Smith, AR—OK	706	139	43.6	483
Fort Wayne, IN	1402	103	31.5	323
Fresno, CA.................	251	33.8	23.5	136
Gadsden, AL	260	73	40.9	410
Gainesville, FL.............	186	42.1	19.8	309
Galveston—Texas City, TX	236	56.4	31.5	202
Gary—Hammond— East Chicago, IN	378	83.9	39.9	109

*All odds are presented excluding – 1.

Standard Metropolitan Statistical Area	Crime Index total	Murder and non-negligent man-slaughter	Forcible rape	Robbery
Grand Forks, ND—MN .... Includes Grand Forks County, ND and Polk County, MN	29.9	19999	8332	11110
Grand Rapids, MI .... Includes Kent and Ottawa Counties	21.6	32257	2659	1124
Great Falls, MT .... Includes Cascade County	13.8	17240	7812	1191
Greeley, CO .... Includes Weld County	17.1	12194	2808	1886
Green Bay, WI .... Includes Brown County	27.4		11627	4853
Greensboro—Winston-Salem—High Point, NC .... Includes Davidson, Forsyth, Guilford, Randolph, Stokes and Yadkin Counties	20.7	10525	4629	1141
Greenville—Spartanburg, SC .. Includes Greenville, Pickens and Spartanburg Counties	17.9	7462	2923	1106
Hamilton—Middletown, OH ... Includes Butler County	17.0	14705	4148	1024
Harrisburg, PA .... Includes Cumberland, Dauphin and Perry Counties	24.2	19999	4114	728
Hartford, CT .... Includes Hartford and Tolland Counties	16.9	19230	4463	461
Honolulu, HA .... Includes Honolulu County	13.0	19230	3875	491
Houston, TX .... Includes Brazoria, Fort Bend, Harris, Liberty, Montgomery and Waller Counties	12.8	4254	1729	296
Huntington—Ashland, WV—KY—OH .... Includes Cabell and Wayne Counties, WV, Boyd and Greenup Counties, KY and Lawrence County, OH	23.4	11764	4291	1308
Huntsville, AL .... Includes Limestone, Madison and Marshall Counties	19.6	9900	3622	1148

Standard Metropolitan Statistical Area	Aggra-vated assault	Burglary	Larceny-theft	Motor vehicle theft
Grand Forks, ND—MN	1817	228	37.5	451
Grand Rapids, MI	522	83.8	35.9	447
Great Falls, MT	287	65.9	22.9	166
Greeley, CO	319	64.5	29.9	291
Green Bay, WI	2769	150.0	36.9	650
Greensboro—Winston-Salem—High Point, NC	305	75.2	37.2	435
Greenville—Spartanburg, SC ..	249	68.1	32.5	311
Hamilton—Middletown, OH ...	515	64.5	27.7	429
Harrisburg, PA	583	88.7	42.6	478
Hartford, CT	541	70.9	33.1	121
Honolulu, HA	2095	51.3	22.0	164
Houston, TX	654	47.3	26.4	108
Huntington—Ashland, WV—KY—OH	410	98.6	39.6	379
Huntsville, AL	436	58.6	40.1	262

*All odds are presented excluding − 1.

Standard Metropolitan Statistical Area	Crime Index total	Murder and non-negligent man-slaughter	Forcible rape	Robbery
Indianapolis, IN . . . . Includes Boone, Hamilton, Hancock, Hendricks, Johnson, Marion, Morgan and Shelby Counties	16.8	11493	2386	447
Jackson, MI . . . . Includes Jackson County	17.1	14705	3288	1232
Jackson, MS . . . . Includes Hinds and Rankin Counties	20.0	6896	3520	586
Jacksonville, FL . . . . Includes Baker, Clay, Duval, Nassau, and St. Johns Counties	14.3	7298	2082	457
Janesville—Beloit, WI . . . . Includes Rock County	20.5		8695	1956
Jersey City, NJ . . . . Includes Hudson County	16.0	12499	3389	355
Johnson City—Kingsport—Bristol, TN—VA . . . . Includes Carter, Hawkins, Sullivan, Unicoi, and Washington Counties, TN, Bristol City and Scott and Washington Counties, VA	44.1	22726	11764	2923
Johnstown, PA . . . . Includes Cambria and Somerset Counties	58.0	62499	17856	2409
Kalamazoo—Portage, MI . . . . . Includes Kalamazoo and Van Buren Counties	14.5	18867	3649	899
Kansas City, MO—KA . . . . Includes Cass, Clay, Jackson, Platte and Ray Counties, MO and Johnson and Wyandotte Counties, KA	15.0	7936	2211	442
Kenosha, WI . . . . Includes Kenosha County	14.3	24999	2694	752
Killeen—Temple, TX . . . . Includes Bell and Coryell Counties	22.3	12499	2251	983
Knoxville, TN . . . . Includes Anderson, Blount, Knox and Union Counties	23.3	15151	4015	791

Standard Metropolitan Statistical Area	Aggra- vated assault	Burglary	Larceny- theft	Motor vehicle theft
Indianapolis, IN	525	70.3	31.0	168
Jackson, MI	250	64.9	30.3	368
Jackson, MS	589	61.6	38.9	365
Jacksonville, FL	211	52.5	27.0	299
Janesville—Beloit, WI	1655	111.0	27.8	606
Jersey City, NJ	438	55.5	46.5	84
Johnson City—Kingsport— Bristol, TN—VA	1099	138.0	83.4	563
Johnstown, PA	852	186.0	114.0	838
Kalamazoo—Portage, MI	226	61.3	24.0	403
Kansas City, MO—KA	246	53.2	29.8	221
Kenosha, WI	1452	66.7	23.0	155
Killeen—Temple, TX	706	70.8	40.5	520
Knoxville, TN	470	73.4	49.6	239

*All odds are presented excluding − 1.

Standard Metropolitan Statistical Area	Crime Index total	Murder and non-negligent man-slaughter	Forcible rape	Robbery
La Crosse, WI . . . . Includes La Crosse County	19.7		29411	8064
Lafayette, LA . . . . Includes Lafayette Parish	19.4	11235	4784	1241
Lafayette—West Lafayette, IN . Includes Tippecanoe County	25.5	55555	7575	2277
Lake Charles, LA . . . . Includes Calcasieu Parish	18.8	7873	3095	1221
Lakeland—Winter Haven, FL . . Includes Polk County	12.1	12986	1817	726
Lancaster, PA . . . . Includes Lancaster, PA	38.7	47618	18518	2746
Lansing—East Lansing, MI . . . . Includes Clinton, Eaton, Ingham and Ionia Counties	21.0	24999	3114	1778
Laredo, TX . . . . Includes Webb County	25.2	10637	16948	974
Las Vegas, NV . . . . Includes Clark County	9.6	5649	1476	212
Lawrence, KA . . . . Includes Douglas County	14.5		2251	869
Lawton, OK . . . . Includes Comanche County	19.4	10100	2057	569
Lewiston—Auburn, ME . . . . Includes Androscoggin County	18.5	99999	7936	2221
Lexington—Fayette, KY . . . . . . Includes Bourbon, Clark, Fayette, Jessamine, Scott, and Woodford Counties	17.6	17240	3496	897
Lima, OH . . . . Includes Allen, Auglaize, Putnam and Van Wert Counties	22.9	41666	5024	1526
Lincoln, NE . . . . Includes Lancaster County	18.8	62499	3508	3570
Little Rock—North Little Rock, AR . . . . Includes Pulaski and Saline Counties	13.0	7873	1630	370
Long Branch—Asbury Park, NJ . Includes Monmouth County	18.7	37036	5649	1231

Standard Metropolitan Statistical Area	Aggra- vated assault	Burglary	Larceny- theft	Motor vehicle theft
La Crosse, WI	2267	142.0	25.3	385
Lafeyette, LA..............	283	53.1	42.9	329
Lafeyette—West Lafayette, IN .	1294	91.9	42.7	384
Lake Charles, LA	271	71.5	33.7	350
Lakeland—Winter Haven, FL ..	119	51.2	22.4	240
Lancaster, PA	1847	141.0	63.3	639
Lansing—East Lansing, MI	524	92.2	32.7	488
Laredo, TX	441	98.8	46.3	289
Las Vegas, NV	236	29.6	21.4	140
Lawrence, KA	426	53.0	24.7	332
Lawton, OK	292	66.8	37.6	404
Lewiston—Auburn, ME.......	230	72.4	32.5	382
Lexington—Fayette, KY	490	71.3	28.6	420
Lima, OH	923	89.7	35.5	695
Lincoln, NE	906	98.2	26.5	415
Little Rock—North Little Rock, AR	217	47.5	25.3	206
Long Branch—Asbury Park, NJ .	540	71.3	31.1	341

*All odds are presented excluding – 1.

Standard Metropolitan Statistical Area	Crime Index total	Murder and non-negligent man-slaughter	Forcible rape	Robbery
Longview—Marshall, TX ..... Includes Gregg and Harrison Counties	28.6	4877	3675	1747
Lorain—Elyria, OH .... Includes Lorain County	28.0	19999	5180	879
Los Angeles—Long Beach, CA.. Includes Los Angeles County	12.6	5746	1494	213
Louisville, KY—IN .... Includes Bullitt, Jefferson and Oldham Counties, KY and Clark and Floyd Counties, IN	19.6	12657	2816	520
Lubbock, TX .... Includes Lubbock County	11.7	5290	1596	649
Lynchburg, VA .... Includes Lynchburg City and Amherst, Appomattox and Campbell Counties	25.2	9258	5713	2906
Macon, GA .... Includes Bibb, Houston, Jones and Twiggs Counties	19.5	6060	3952	750
Madison, WI .... Includes Dane County	16.0	83332	4784	1886
Manchester, NH .... Includes Hillsborough County	23.0	249999	9345	2940
Mansfield, OH .... Includes Richland County	19.2	124999	3256	1120
McAllen—Pharr—Edinburg, TX .... Includes Hidalgo County	24.0	14924	8771	4463
Melbourne—Titusville—Cocoa, FL................. Includes Brevard County	14.8	21276	2221	1480
Memphis, TN—AR—MS ...... Includes Shelby and Tipton Counties, TN, Crittendon County, AR and De Soto County, MS	16.9	7352	1254	292
Miami, FL .... Includes Dade County	10.9	5881	2231	223
Midland, TX .... Includes Midland County	27.4	12194	2716	2095

Standard Metropolitan Statistical Area	Aggra- vated assault	Burglary	Larceny– theft	Motor vehicle theft
Longview—Marshall, TX	920	83.6	56.0	426
Lorain—Elyria, OH	482	91.0	58.9	286
Los Angeles—Long Beach, CA..	187	41.8	33.0	100
Louisville, KY—IN...........	766	73.2	36.0	233
Lubbock, TX	310	42.7	21.0	223
Lynchburg, VA	347	127.0	39.1	521
Macon, GA	548	65.6	36.8	267
Madison, WI	1945	74.9	23.0	404
Manchester, NH	1566	81.8	38.8	291
Mansfield, OH	153	84.4	35.7	349
McAllen—Pharr— Edinburg, TX	610	72.8	43.8	522
Melbourne—Titusville— Cocoa, FL.................	297	65.6	23.4	385
Memphis, TN—AR—MS	369	49.5	41.7	175
Miami, FL	129	44.0	22.2	169
Midland, TX	373	84.4	62.1	245

*All odds are presented excluding – 1.

Standard Metropolitan Statistical Area	Crime Index total	Murder and non-negligent man-slaughter	Forcible rape	Robbery
Milwaukee, WI .... Includes Milwaukee, Ozaukee, Washington and Waukesha Counties	21.1	23255	4097	971
Minneapolis-St. Paul, MN—WI . Includes Anoka, Carver, Chicago, Dakota, Hennepin, Ramsey, Scott, Washington and Wright Counties, MN and St. Croix County, WI	17.9	37036	3377	681
Mobile, AL................. Includes Baldwin and Mobile Counties	16.6	5375	2197	458
Modesto, CA .... Includes Stanislaus County	12.7	9523	2631	892
Monroe, LA .... Includes Ouachita Parish	19.0	8695	2777	2293
Montgomery, AL .... Includes Autauga, Elmore and Montgomery Counties	17.5	5154	2777	783
Muskegon—Norton Shores— Muskegon Heights, MI .... Includes Muskegon and Oceana Counties	15.9	16128	2873	1001
Nashville—Davidson, TN ..... Includes Cheatham, Davidson, Dickson, Robertson, Rutherford, Sumner, Williamson and Wilson Counties	21.3	7751	3002	406
Nassau—Suffolk, NY .... Includes Nassau and Suffolk Counties	20.8	52631	11627	1129
Newark, NJ .... Includes Essex, Morris, Somerset and Union Counties	17.0	11363	3076	305
New Brunswick—Perth Amboy —Sayreville, NJ .... Includes Middlesex County	20.2	58823	6210	1035
New Haven—West Haven, CT.. Includes New Haven County	15.9	27777	5617	535
New Orleans, LA .... Includes Jefferson, Orleans, St. Bernard and St. Tammany Parishes	15.6	4201	1937	224

Standard Metropolitan Statistical Area	Aggra-vated assault	Burglary	Larceny-theft	Motor vehicle theft
Milwaukee, WI	1047	118.0	31.4	281
Minneapolis-St. Paul, MN—WI .	819	68.5	31.1	228
Mobile, AL................	237	47.7	37.6	310
Modesto, CA	271	51.9	21.4	232
Monroe, LA	177	113.0	30.1	390
Montgomery, AL	992	57.2	31.3	317
Muskegon—Norton Shores—Muskegon Heights, MI	239	69.2	25.9	445
Nashville—Davidson, TN	481	67.3	45.6	263
Nassau—Suffolk, NY	1134	79.8	35.9	222
Newark, NJ	345	67.2	35.6	141
New Brunswick—Perth Amboy —Sayreville, NJ.............	555	80.7	34.5	252
New Haven—West Haven, CT..	890	55.3	31.1	137
New Orleans, LA	238	56.5	28.1	134

*All odds are presented excluding − 1.

Standard Metropolitan Statistical Area	Crime Index total	Murder and non-negligent man-slaughter	Forcible rape	Robbery
Newport News—Hampton, VA . Includes Hampton, Newport News, Poquoson and Williamsburg Cities and Gloucester, James City, and York Counties	21.9	13888	2603	888
New York, NY—NJ .... Includes Bronx, Kings, New York, Putnam, Queens, Richmond, Rockland and Westchester Counties, NY and Bergen County, NJ	13.1	5951	2282	121
Norfolk—Virginia Beach—Portsmouth, VA—NC .... Includes Chesapeake, Norfolk, Portsmouth, Suffolk, and Virginia Beach Cities, VA, and Currituck County, NC	16.8	10308	2468	557
North East, PA .... Includes Lackawanna, Luzerne Monroe Counties	36.2	28570	10637	2702
Odessa, TX .... Includes Ector County	17.2	6368	4607	1223
Oklahoma City, OK .... Includes Canadian, Cleveland, McClain, Oklahoma and Pottwatomie Counties	16.6	9258	1983	630
Omaha, NE .... Includes Douglas and Sarpy Counties, NE and Pottawatomie County, IA	17.5	18181	2914	577
Orlando, FL .... Includes Orange, Osceola and Seminole Counties	10.9	9008	1620	533
Owensboro, KY .... Includes Daviess County	24.5	20407	7352	1723
Oxnard—Simi Valley—Ventura, CA .... Includes Ventura County	17.7	15151	2754	660
Panama City, FL .... Includes Bay County	18.3	15384	3558	872

Standard Metropolitan Statistical Area	Aggra-vated assault	Burglary	Larceny-theft	Motor vehicle theft
Newport News—Hampton, VA .	529	86.5	36.7	457
New York, NY—NJ	203	48.9	35.7	100
Norfolk—Virginia Beach—Portsmouth, VA—NC	360	76.6	27.4	321
North East, PA..............	1343	130.0	63.3	410
Odessa, TX	577	78.9	27.2	248
Oklahoma City, OK	270	50.8	37.4	180
Omaha, NE	485	75.5	29.7	233
Orlando, FL................	156	39.8	20.4	240
Owensboro, KY	716	89.3	42.2	347
Oxnard—Simi Valley—Ventura, CA	420	55.3	36.5	219
Panama City, FL	444	65.5	32.8	296

*All odds are presented excluding − 1.

Standard Metropolitan Statistical Area	Crime Index total	Murder and non-negligent man-slaughter	Forcible rape	Robbery
Parkersburg—Marietta, WV— OH...................... Includes Wirt and Wood Counties, WV and Washington County, OH	26.0	52631	5713	2403
Pascagoula—Moss Point, MS .. Includes Jackson County	22.0	9523	3583	924
Paterson—Clifton— Passaic, NJ .... Includes Passaic County	17.3	13888	9173	443
Pensacola, FL .... Includes Escambia and Santa Rosa Counties	12.5	7691	1694	585
Peoria, IL.................. Includes Peria, Tazewell and Woodford Counties	19.7	35713	3703	4920
Petersburg—Colonial Heights— Hopewell, VA .... Includes Colonial Heights, Hopewell and Petersburg Cities and Dinwiddie and Prince George Counties	21.2	6943	4166	843
Philadelphia, PA—NJ .... Includes Bucks, Chester, Delaware, Montgomery and Philadelphia Counties, PA and Burlington, Camden and Gloucester Counties, NJ	23.2	10203	3662	468
Phoenix, AZ................ Includes Maricopa County	10.5	9999	1889	478
Pine Bluff, AR .... Includes Jefferson County	20.1	6172	1922	714
Pittsburgh, PA .... Includes Allegheny, Beaver, Washington and Westmoreland Counties	32.8	19230	4999	562
Pittsfield, MA .... Includes Berkshire County	22.4	142856	11363	1584
Portland, ME .... Includes Cumberland County	15.0	41666	6943	1519
Portland, OR—WA .... Includes Clackamas, Multnomah and Washington Counties, OR and Clark County, WA	13.5	21276	1861	464

Standard Metropolitan Statistical Area	Aggravated assault	Burglary	Larcenytheft	Motor vehicle theft
Parkersburg—Marietta, WV—OH......................	840	92.6	43.9	452
Pascagoula—Moss Point, MS ..	415	50.6	57.2	321
Paterson—Clifton—Passaic, NJ	404	65.3	35.7	136
Pensacola, FL	212	47.1	23.2	207
Peoria, IL..................	218	78.8	35.2	459
Petersburg—Colonial Heights—Hopewell, VA	564	93.4	33.2	509
Philadelphia, PA—NJ	492	87.7	47.4	208
Phoenix, AZ................	285	41.1	18.3	186
Pine Bluff, AR	438	53.5	42.8	511
Pittsburgh, PA	714	120.0	74.2	215
Pittsfield, MA	586	61.4	45.4	367
Portland, ME	366	55.9	25.9	235
Portland, OR—WA	295	50.2	25.2	204

*All odds are presented excluding − 1.

Standard Metropolitan Statistical Area	Crime Index total	Murder and non-negligent man-slaughter	Forcible rape	Robbery
Poughkeepsie, NY .... Includes Dutchess County	26.4	17543	7936	1489
Providence—Warwick—Pawtucket, RI .... Includes Bristol, Kent, Providence and Washington Counties	19.2	23809	9173	987
Provo—Orem, UT .... Includes Utah County	26.9	62499	11493	6368
Pueblo, CA................. Includes Pueblo County	15.0	24389	2325	1053
Racine, WI................. Includes Racine County	16.8	10525	4236	837
Raleigh—Durham, NC........ Includes Durham, Orange and Wake Counties	19.0	10416	3085	1155
Rapid City, SD .... Includes Meade and Pennington Counties	16.9	9091	4064	1861
Reading, PA................ Includes Berks County	32.0	21276	7873	1088
Reno, NV .... Includes Washoe County	10.6	6368	1945	301
Richland—Kennewick—Pasco, WA................. Includes Benton and Franklin Counties	19.0	24389	2746	1390
Richmond, VA .... Includes Richmond City and Charles City, Chesterfield, Goochland, Hanover, Henrico New Kent and Powhatan Counties	17.5	8064	3134	599
Riverside—San Bernardino—Ontario, CA .... Includes Riverside and San Bernardino Counties	12.7	7575	1872	474
Roanoke, VA .... Includes Roanoke and Salem Cities and Botetourt, Craig and Roanoke Counties	17.5	10637	5434	1003
Rochester, MN.............. Includes Olmsted County	23.6		3400	2182

Standard Metropolitan Statistical Area	Aggra-vated assault	Burglary	Larceny-theft	Motor vehicle theft
Poughkeepsie, NY	483	87.4	47.4	628
Providence—Warwick—Pawtucket, RI	438	72.2	36.8	119
Provo—Orem, UT	991	169.0	36.4	531
Pueblo, CA.................	148	65.2	26.8	300
Racine, WI.................	456	71.6	26.8	382
Raleigh—Durham, NC	432	71.6	32.3	384
Rapid City, SD	205	70.4	29.8	274
Reading, PA................	885	94.2	63.8	474
Reno, NV	511	40.6	18.9	168
Richland—Kennewick—Pasco, WA.................	430	83.4	30.2	370
Richmond, VA	524	69.3	29.0	287
Riverside—San Bernardino—Ontario, CA	243	40.0	26.5	185
Roanoke, VA	759	71.8	27.2	434
Rochester, MN..............	1638	104.0	35.7	393

*All odds are presented excluding − 1.

Standard Metropolitan Statistical Area	Crime Index total	Murder and non-negligent man-slaughter	Forcible rape	Robbery
Rochester, NY .... Includes Livingston, Monroe, Ontario, Orleans and Wayne Counties	17.2	20407	5779	813
Rockford, IL .... Includes Boone and Winnebago Counties	17.2	22726	3801	720
Sacramento, CA .... Includes Placer, Sacramento and Yolo Counties	11.0	8695	1987	322
Saginaw, MI .... Includes Saginaw County	14.7	7041	2015	577
St. Cloud, MN .... Includes Benton, Sherburne and Stearns Counties	42.3	38467	11904	7751
St. Joseph, MO .... Includes Andrew and Buchanan Counties	16.4	24999	6288	1257
St. Louis, MO—IL .... Includes St. Louis City and Franklin, Jefferson, St. Charles and St. Louis Counties, MO and Clinton, Madison, Monroe and St. Clair Counties, IL	17.2	6943	2450	324
Salem, OR .... Includes Marion and Polk Counties	16.2	19230	2409	1411
Salinas—Seaside—Monterey, CA .... Includes Monterey County	17.9	16128	2590	700
Salt Lake City—Ogden, UT .... Includes Davis, Salt Lake, Tooele and Weber Counties	15.4	25640	3377	1042
San Angelo, TX .... Includes Tom Green County	15.4	8849	3174	1844
San Antonio, TX .... Includes Bexar, Comal and Guadalupe Counties	15.8	6710	2761	650
San Diego, CA .... Includes San Diego County	13.8	14705	2731	446
San Francisco—Oakland, CA .. Includes Alameda, Contra Costa, Marin, San Francisco and San Mateo Counties	11.4	8928	1886	240

Standard Metropolitan Statistical Area	Aggra-vated assault	Burglary	Larceny-theft	Motor vehicle theft
Rochester, NY	500	69.2	28.3	321
Rockford, IL	402	65.1	29.5	352
Sacramento, CA	279	40.5	21.2	150
Saginaw, MI	236	59.7	25.1	414
St. Cloud, MN	4544	235.0	57.9	706
St. Joseph, MO	698	61.1	26.8	355
St. Louis, MO—IL	298	59.3	33.7	180
Salem, OR	413	65.7	26.6	323
Salinas—Seaside—Monterey, CA	464	62.9	32.3	328
Salt Lake City—Ogden, UT	456	64.0	25.3	236
San Angelo, TX	660	76.6	22.7	283
San Antonio, TX	539	51.8	29.6	252
San Diego, CA	424	47.1	27.4	151
San Francisco—Oakland, CA ..	292	43.5	22.4	142

*All odds are presented excluding − 1.

Standard Metropolitan Statistical Area	Crime Index total	Murder and non-negligent man-slaughter	Forcible rape	Robbery
San Jose, CA .... Includes Santa Clara County	14.5	16392	2074	640
Santa Barbara—Santa Maria—Lompoc, CA .... Includes Santa Barbara County	14.3	29411	3011	1137
Santa Cruz, CA .... Includes Santa Cruz County	14.0	28570	2450	827
Santa Rosa, CA .... Includes Sonoma County	14.8	15872	2949	926
Sarasota, FL .... Includes Sarasota County	17.1	34482	2570	2627
Savannah, GA .... Includes Bryan, Chatham and Effingham Counties	12.9	6172	1596	465
Seattle—Everett, WA .... Includes King and Snohomish Counties	13.5	17240	2019	506
Sherman—Denison, TX .... Includes Grayson County	26.5	14705	8771	2444
Shreveport, LA .... Includes Bossier, Caddo and Webster Parishes	16.7	5847	2374	837
Sioux City, IA—NB .... Includes Woodbury County, IA, and Dakota County, NB	18.4	62499	4347	2095
Sioux Falls, SD............. Includes Minnehaha County	22.9	25640	5347	3076
South Bend, IN .... Includes Marshall and St. Joseph Counties	19.2	18518	3154	670
Spokane, WA............... Includes Spokane County	16.5	18867	3447	882
Springfield, IL .... Includes Menard and Sangamon Counties	11.7	9708	4097	414
Springfield, MO............. Includes Christian and Greene Counties	13.7	20832	4291	1026
Springfield, OH .... Includes Champaign and Clark Counties	24.1	37036	7575	748

Standard Metropolitan Statistical Area	Aggra-vated assault	Burglary	Larceny-theft	Motor vehicle theft
San Jose, CA	513	50.1	26.9	192
Santa Barbara—Santa Maria—Lompoc, CA	374	55.8	23.5	314
Santa Cruz, CA	373	52.8	24.6	207
Santa Rosa, CA	497	51.9	26.5	214
Sarasota, FL	283	66.0	26.5	412
Savannah, GA	154	49.9	24.6	259
Seattle—Everett, WA	357	48.9	25.1	197
Sherman—Denison, TX	977	88.9	45.9	424
Shreveport, LA	308	60.6	30.2	328
Sioux City, IA—NB	1020	97.3	26.7	275
Sioux Falls, SD	1034	106.0	33.97	420
South Bend, IN	1057	67.7	32.98	377
Spokane, WA	537	66.2	27.5	259
Springfield, IL	316	323.0	25.0	236
Springfield, MO	536	51.7	22.9	263
Springfield, OH	1116	94.5	38.7	558

*All odds are presented excluding − 1.

Standard Metropolitan Statistical Area	Crime Index total	Murder and non- negligent man- slaughter	Forcible rape	Robbery
Steubenville—Weirton, OH— WV .... Includes Jefferson County, OH, and Brooke and Hancock Counties, WV	45.0	12820	13888	1035
Stockton, CA .... Includes San Joaquin County	11.7	5713	1772	368
Syracuse, NY.............. Includes Madison, Onondaga and Oswego Counties	18.7	38461	7812	862
Tacoma, WA .... Includes Pierce County	15.2	22221	1694	663
Tallahassee, FL .... Includes Leon and Wakulla Counties	11.9	13698	1331	757
Tampa—St. Petersburg, FL .... Includes Hillsborough, Pasco and Pinellas Counties	13.7	11904	1858	569
Terre Haute, IN .... Includes Clay, Sullivan, Vermillion and Vigo Counties.	22.2	21738	5987	1657
Texarkana, TX—Texarkana, AR Includes Bowie County, TX, and Little River and Miller Counties, AR	23.7	8474	3967	1564
Toledo, OH—MI .... Includes Fulton, Lucas, Ottawa and Wood Counties, Ohio and Monroe County, MI	16.1	13513	2694	451
Topeka, KS .... Includes Jefferson, Osage and Shawnee Counties	18.1	12986	3002	955
Trenton, NJ .... Includes Mercer County	14.6	28570	3095	334
Tucson, AZ .... Includes Pima County	11.0	12345	2187	602
Tulsa, OK .... Includes Creek, Mayes, Osage, Rogers, Tulsa and Wagoner Counties	18.0	10416	2563	906
Tuscaloosa, AL .... Includes Tuscaloosa County	17.6	17856	2906	1058

Standard Metropolitan Statistical Area	Aggravated assault	Burglary	Larceny–theft	Motor vehicle theft
Steuvenville—Weirton, OH—WV	580	132.0	102.0	612
Stockton, CA	308	42.7	22.8	140
Syracuse, NY..............	1118	64.0	31.9	353
Tacoma, WA	446	46.0	30.1	259
Tallahassee, FL	193	45.8	20.9	310
Tampa—St. Petersburg, FL	214	49.6	25.2	331
Terre Haute, IN	1512	81.1	38.9	229
Texerkana, TX—Texarkana, AR	408	124.0	36.7	397
Toledo, OH—MI	493	67.7	26.9	284
Topeka, KS	409	69.9	29.9	509
Trenton, NJ	490	52.8	27.8	191
Tucson, AZ	356	40.7	19.3	188
Tulsa, OK	376	62.8	34.9	208
Tuscaloosa, AL	249	64.9	32.7	277

*All odds are presented excluding – 1.

Standard Metropolitan Statistical Area	Crime Index total	Murder and non-negligent man-slaughter	Forcible rape	Robbery
Tyler, TX .... Includes Smith County	15.5	6578	3730	741
Utica—Rome, NY .... Includes Herkimer and Oneida Counties	35.6	39499	21738	1937
Vallejo—Fairfield—Napa, CA . Includes Napa and Solano Counties	15.8	17856	2881	853
Vineland—Millville—Bridgeton, NJ .... Includes Cumberland County	16.7	32257	3343	911
Waco, TX .... Includes McLennan County	17.0	7633	2906	897
Washington, DC—MD—VA ... Includes District of Columbia, Charles, Montgomery, and Prince Georges Counties, MD, Alexandria, Fairfax, Falls Church, Manassas, and Manassas Park Cities, and Arlington, Fairfax, Loudoun, and Prince William Counties, VA	15.7	10308	2524	280
Waterloo—Cedar Falls, IA .... Includes Black Hawk County	15.6	142856	7999	961
West Palm Beach—Boca Raton, FL .... Includes Palm Beach County	11.4	9708	2414	506
Wheeling, WV—OH .... Includes Marshall and Ohio Counties, WV and Belmont County, OH	41.5	37036	6802	1999
Wichita, KS .... Includes Butler and Sedgwick Counties	14.9	12047	2397	665
Wichita Falls, TX............ Includes Clay and Wichita Counties	17.8	9614	2792	585
Williamsport, PA............ Includes Lycoming County	27.3	38461	5987	2140
Wilmington, DL—NJ—MD .... Includes New Castle County, DL, Salem County, NJ, and Cecil County, MD	15.1	19230	4404	652

Standard Metropolitan Statistical Area	Aggravated assault	Burglary	Larceny-theft	Motor vehicle theft
Tyler, TX	823	57.1	25.9	295
Utica—Rome, NY	1685	93.8	69.4	681
Vallejo—Fairfield—Napa, CA .	281	62.2	27.2	313
Vineland—Millville—Bridgeton, NJ	600	55.4	30.2	275
Waco, TX	219	69.1	29.8	357
Washington, DC—MD—VA . . .	458	69.3	27.4	239
Waterloo—Cedar Falls, IA	748	90.8	21.8	347
West Palm Beach—Boca Raton, FL	184	42.5	20.8	239
Wheeling, WV—OH	1308	162.0	70.5	504
Wichita, KS	477	56.8	25.6	254
Wichita Falls, TX	351	73.9	30.6	317
Williamsport, PA	859	103.0	44.0	572
Wilmington, DL—NJ—MD	482	60.6	26.3	187

*All odds are presented excluding – 1.

Standard Metropolitan Statistical Area	Crime Index total	Murder and non-negligent man-slaughter	Forcible rape	Robbery
Wilmington, NC . . . . Includes Brunswick and New Hanover Counties	14.9	11904	3288	848
Worcester, MA . . . . Includes Worcester County	22.9	71428	6802	117
Yakima, WA . . . . Includes Yakima County	13.1	19607	2898	759
York, PA . . . . Includes Adams and York Counties	27.7	22726	6666	899
Youngstown—Warren, OH . . . . Includes Mahoning and Trumbull Counties	24.6	16948	4694	810

Standard Metropolitan Statistical Area	Aggra-vated assault	Burglary	Larceny-theft	Motor vehicle theft
Wilmington, NC	237	50.7	28.0	293
Worcester, MA	624	75.6	53.4	129
Yakima, WA	264	52.0	22.7	239
York, PA	934	99.7	48.3	451
Youngstown—Warren, OH	540	96.1	43.7	329

*All odds are presented excluding – 1.

Chapter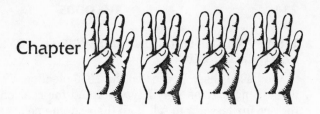

BY THE PEOPLE, FOR THE PEOPLE
Citizenship, Elections and Government

As you'll see in the pages which follow, our government isn't nearly representative enough. Fewer Americans are registering and voting than ever before.

This pattern is far from uniform among different population groups. Home-owners, farmers, midwesterners, the better educated and whites are better than average citizens. Southerners, the unemployed and those of Spanish origin are poor at turning out to vote. And if parents do set an example for their kids we're in trouble: childless couples are far more apt to go to the polls than those with children.

This chapter shows that when election time rolls round it's better to be a Democratic congressman than a Republican senator. And speaking of the Senate, we give you a state by state rundown on the odds of entering that august body.

Next we examine the legislative process. The odds of a bill being passed are long. If passed the president's not apt to veto it but if he does the veto generally sticks.

Finally, we examine presidential mortality. Given the morbid picture here it's a wonder anyone wants the job.

WHAT ARE THE ODDS ON A MEMBER OF CONGRESS OR SENATOR BEING REELECTED?

We based these figures on the 1976 election. They clearly show that while congressmen may have to stand for reelection every two years, they're far less apt to fall than their senate colleagues.

The Odds: Congressman   **22.8-1** for
            Senator   **1.8-1** for

Source: Congressional Research Service

WHAT ARE THE ODDS ON VOTING BY AREA OF EMPLOYMENT?

At 81% self-improved agricultural workers were the leading category in terms of voter participation in 1978. Government workers at 77.2% were next. Given the government is their field of interest (and that sometimes their job depends on the outcome of the vote), this is understandable. The lowest categories in terms of participation were farm employees at 43.5% and the unemployed at 44.1%.

The Odds: for unless noted

Agriculture
Self-employed workers   4.3-1
Wage and salary workers   1.3-1 against

Non-agricultural industries
Private wage and salary workers   1.4-1
Government workers   3.4-1
Self-employed workers   2.3-1

Unemployed   1.3-1 against

Source: Department of Commerce, Bureau of the Census, *Voting and Registration in the Elections of November 1978*, Series P-20, No. 344, 1979, p. 65.

WHAT ARE THE ODDS ON THE HEAD OF THE FAMILY VOTING, IN TERMS OF HOME OWNERSHIP?

One of two primary heads of a family voted in the 1978 elections. Home-owners clearly feel far more strongly about voting (58.5% did so) than renters.

The Odds: Total   **Even**
            Home-owners   **1.4-1** for
            Renters   **2.6-1** against

Source: Ibid., p. 2.

WHAT ARE THE ODDS THAT THOSE WITH AND WITHOUT CHILDREN WILL VOTE?

You'd think parents would be more responsible citizens. The exact opposite is true. In 1978 among primary heads of families 56.3% voted, compared to 44.9% who had children.

The Odds:	All heads of families	**Even**	
	Those with children	**1.2-1** against	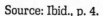
	Those without children	**1.3-1** for	

Source: Ibid., p. 4.

WHAT ARE THE ODDS ON VOTING BASED ON EDUCATION LEVEL?

In 1978 just 28.7% of those who hadn't finished grade school voted, compared to 63.9% of college graduates.

The Odds: against unless noted

Education level		
Less than 8 years	2.5-1	
8 years	1.4-1	
Some high school	1.9-1	
Finished high school	1.2-1	
1-3 years college	1.1-1 for	
4 or more years college	2.5-1 for	

Source: Ibid., p. 4.

WHAT ARE THE ODDS BY RACE ON VOTING IN A PRESIDENTIAL OR CONGRESSIONAL ELECTION?

Presidential elections attract more voters than those held in mid-term (recently about 25% more). Thus 59.2% of those of voting age cast a ballot in 1976 while only 45.9% did so in 1978.

Compared to many countries these aren't particularly good citizen participation rates. What's perhaps more alarming is that proportionately fewer Americans vote with each passing four year election cycle.

Whites are most apt to exercise their suffrage while those of Spanish origin are least likely to do so. In spite of gains in black voter registrations, blacks vote less often than whites (37.2% did so in 1978 versus 47.3% for whites). Along with other population groups, a smaller percentage of those eligible have bothered to vote in recent years.

The Odds: against unless noted

	1978 congressional election	1976 presidential election	
Total	1.2-1	1.5-1 for	
White	1.1-1	1.6-1 for	
Black	1.7-1	Even	
Spanish	3.2-1	2.1-1 against	

Source: Department of Commerce, Bureau of the Census, *Voting and Registration in the Elections of November 1978,* Series P-20, No. 344, 1979, p. 2.

WHAT ARE THE ODDS ON A CONGRESSMAN RUNNING UNOPPOSED BY PARTY?

Based on the 1974, '76 and '78 elections an average of 52 Democrats ran unopposed, compared to 8.6 Republicans.

The Odds: Democrat   **4.6-1** against
                Republican   **16.5-1** against

Source: *Congressional Quarterly Weekly Report,* No. 13, March 1979.

WHAT ARE THE ODDS BY STATE ON BECOMING A SENATOR?

If you have a hankering to be a senator head for Alaska, Wyoming or Vermont. In all three you'll be over 35 times more certain of being selected to run than in New York, where senators have a constituency of nearly 19 million.

The Odds: against

New England
Maine	545,000-1	
New Hampshire	435,500-1
Vermont	243,500-1
Massachusetts	1,887,000-1
Rhode Island	467,500-1
Connecticut	1,549,500-1

Middle Atlantic
New York	8,874,000-1
New Jersey	3,663,000-1
Pennsylvania	5,875,000-1

East North Central
Ohio	5,374,500-1
Indiana	2,687,000-1

The Odds: against

Illinois	5,621,500-1
Michigan	4,594,500-1
Wisconsin	2,339,500-1

West North Central

Minnesota	2,004,000-1
Iowa	1,448,000-1
Missouri	2,430,000-1
North Dakota	326,000-1
South Dakota	345,000-1
Nebraska	782,000-1
Kansas	1,174,000-1

South Atlantic

Delaware	291,500-1
Maryland	2,071,500-1
Virginia	2,574,000-1
West Virginia	930,000-1
North Carolina	2,788,500-1
South Carolina	1,459,000-1
Georgia	2,542,000-1
Florida	4,297,000-1

East South Central

Kentucky	1,749,000-1
Tennessee	2,178,500-1
Alabama	1,871,000-1
Mississippi	1,201,000-1

West South Central

Arkansas	1,093,000-1
Louisiana	1,983,000-1
Oklahoma	1,440,000-1
Texas	6,507,000-1

Mountain

Montana	392,500-1
Idaho	439,000-1
Wyoming	212,000-1
Colorado	1,335,000-1
New Mexico	606,000-1
Arizona	1,177,000-1
Utah	653,500-1
Nevada	330,000-1

The Odds: against

	Pacific
Washington	1,887,000-1
Oregon	1,222,000-1
California	11,147,000-1
Alaska	201,500-1
Hawaii	448,500-1

Source: Data from 1978 Census.

WHAT ARE THE ODDS ON A SENATOR HAVING BEEN A REPRESENTATIVE?

Thirty-one of our current senators served in the House.

The Odds: **2.2-1** against

Source: 1979 *Congressional Directory*, 96th Congress, 1st Session, pp. 239-241.

WHAT ARE THE ODDS ON ENTERING THE SENATE WITHOUT HOLDING PRIOR PUBLIC OFFICE?

Eighteen of our 100 senators entered the Senate without serving in a prior elective office.

The Odds: **4.6-1** against

Source: Congressional Research Service

WHAT ARE THE ODDS OF A BILL PASSING IN THE SENATE OR THE HOUSE?

We based our figures on an analysis of the 50,474 bills introduced in the House and the 13,803 bills introduced in the Senate during the 93rd through the 95th Congresses.

The Odds: Senate   18.6-1 against
           House   42.7-1 against

Source: "U.S. Congress Calendars of the U.S. House of Representatives," *History of Legislation Annual*.

WHAT ARE THE ODDS ON A PRESIDENTIAL VETO?

We analyzed figures from 1969 through 1977. Of the 3,460 measures enacted the president vetoed 116.

The Odds: **28.8-1** against

Source: *Congressional Research Service*

WHAT ARE THE ODDS ON A PRESIDENT'S VETO BEING OVERRIDDEN?

Twenty-eight of the 116 bills were passed over the veto.

The Odds: **3.1-1** against

Source: Ibid.

WHAT ARE THE ODDS ON A PRESIDENT HAVING BEEN A SENATOR?

Fourteen of our 39 presidents were previously senators. This tradition started with our fifth president, John Quincy Adams, in 1824. The last president with prior senatorial experience was Richard Nixon.

The Odds: **1.8-1** against

Source: Ibid.

WHAT ARE THE ODDS ON A PRESIDENT DYING IN OFFICE?

In all, nine of our 39 presidents have died in office. Since Lincoln's assassination in 1865, four of our leaders have been killed while in office.

The Odds: **3.3-1** against

Source: Ibid.

WHAT ARE THE ODDS ON A PRESIDENT ELECTED IN A YEAR ENDING IN "0" DYING IN OFFICE?

Since 1800 only two have survived, Jefferson and Monroe. Since 1840 none have lived out their term.

The Odds: **3.5-1** for

Source: Elections Research Center, D.C.

WHAT ARE THE ODDS ON BEING REGISTERED AND VOTING IN OUR 25 MOST POPULOUS STATES?

Good citizenship is a way of life in Minnesota and Wisconsin. In 1978 83.5% of eligible Minnesotians were registered and 64.2% voted. In Wisconsin 85.3% are registered and 58.3% voted. By way of contrast, just 29% of Georgians and 29.7% of Kentuckians voted.

The Odds: for unless noted

	Registered	Voting	
California	1.4-1	1.1-1	against
New York	1.4-1	1.2-1	against
Texas	1.2-1	1.9-1	against
Pennsylvania	1.6-1	Even	
Illinois	1.9-1	1.1-1	against
Ohio	1.5-1	1.2-1	against
Michigan	2.3-1	1.2-1	
Florida	1.5-1	1.1-1	against
New Jersey	1.6-1	1.4-1	against
Massachusetts	2.3-1	1.3-1	
North Carolina	Even	2.2-1	against
Indiana	1.6-1	1.3-1	against
Virginia	1.3-1	1.3-1	against
Missouri	2.3-1	1.1-1	
Georgia	1.4-1	2.4-1	against
Wisconsin	5.8-1	1.4-1	
Tennessee	1.7-1	1.2-1	against
Maryland	1.7-1	1.2-1	against
Minnesota	5.1-1	1.8-1	
Washington	1.5-1	1.5-1	against
Louisiana	2.1-1	1.7-1	against
Alabama	2.1-1	1.1-1	against
Kentucky	1.6-1	2.4-1	against
Connecticut	2.1-1	1.2-1	
Iowa	1.8-1	1.2-1	against

Source: Voting & Registration in the Elections of November 1978, pp. 30-31.

Chapter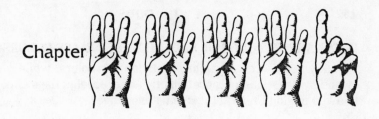

ODDS AND ENDS

WHAT ARE THE ODDS OF SAILING UNDER FALSE COLORS?

Today 500 merchant ships fly under our colors, while another 700 owned by American corporations fly under foreign flags of convenience to avoid paying United States maritime wages.

The Odds: **1.4-1** for

Source: U.S. Maritime Administration

WHAT ARE THE ODDS OF GOING TO THE DOGS?

The American Greyhound Track Operators Association reports 1978 track attendance at 20,046,298. We assumed people went twice on the average.

The Odds: **20.9-1** against

Source: Data from American Greyhound Track Operators Association.

WHAT ARE THE ODDS THAT SOMEONE'S FOR THE BIRDS?

No one's more for the birds than the dedicated 100 employees of Ducks Unlimited in Chicago.

The Odds: **2,195,299-1** against

Source: Ducks Unlimited

WHAT ARE THE ODDS ON PLAYING SECOND FIDDLE?

The New York Philharmonic has 106 orchestra members, 18 of whom play second violin, in other words second fiddle. The Cleveland Symphony Orchestra has 103 musicians with 15 second violinists, while for Philadelphia the comparable figures are 102 and 16. So, approximately, you're more apt to play second fiddle in New York.

The Odds: **5.2-1** against

Source: Data from various orchestras

WHAT ARE THE ODDS THAT "THAT DOG WON'T HUNT"?

Of the 122 breeds recognized by the American Kennel Club just 24 are classified as sporting breeds.

The Odds: **4.1-1** for

Source: American Kennel Club

WHAT ARE THE ODDS OF BRINGING HOME THE BACON THIS WEEK?

Since over one and a half billion pounds of bacon are consumed each year by the 90% of all families who eat it, the odds (notably in the South) are good.

The Odds: **1.5-1** for

Source: Data from Armour Foods, Greyhound Corporation, Phoenix, Arizona.

WHAT ARE THE ODDS ON BEING TOP BANANA?

The average banana plant (contrary to popular belief they're not trees) produces one stem with an average of 100 bananas. One is invariably top of the clump.

The Odds: **99.0-1** against

Source: Data from United Fruit Company, New York City.

WHAT ARE THE ODDS ON DELIVERING THE GOODS?

In 1977 4,183,000 people worked in trucking, on the railroads or in commercial aviation. We used all those over 18 that year as our base.

The Odds: **39.3-1** against

Source: Bureau of Labor

WHAT ARE THE ODDS OF COMING ON LIKE GANGBUSTERS?

The Justice Department reports that there are currently 16 Organized Crime Task Forces across the nation with total manpower of 192.

The Odds: **1,142,187-1** against

Source: Justice Department

WHAT ARE THE ODDS OF BEING A PHILADELPHIA LAWYER?

There are 464,851 lawyers in the United States and 6,900 lawyers in Philadelphia.

The Odds: **66,369-1** against

Source: American Bar Association and the Philadelphia Bar Association.

WHAT ARE THE ODDS ON BEING CAUGHT WITH YOUR HAND IN THE TILL?

In 1977 some 14,542,000 Americans worked in retailing and thus constantly had their hands in the till.

The Odds: **14.1-1** against

Source: Data from Bureau of Labor Statistics.

WHAT ARE THE ODDS THAT IT'S A TURKEY?

In 1978 of the 3,746,300,000 domestic fowl raised, 140 million were turkeys.

The Odds: **25.7-1** against

Source: Data from U.S. Department of Agriculture.

WHAT ARE THE ODDS OF LOOKING AT THE WORLD THROUGH ROSE-TINTED GLASSES?

About 20% of all Americans over the age of 12 wear tinted glasses. A check of 20 optometrists reveals that rose colors account for about 6% of this.

The Odds: **82.3-1** against

Source: Data from American Optometric Association and field research.

WHAT ARE THE ODDS THAT IT'S JUST WHAT THE DOCTOR ORDERED?

A government study shows that about 140 million of America's 1.4 billion prescriptions are mislabeled.

The Odds: **9.0-1** against

Source: Data from Food and Drug Administration.

WHAT ARE THE ODDS ON BEING TOP DOG?

In 1977 there were 775 all-breed dog shows with a grand total of 894,009 dogs competing for best of show.

The Odds: **1,153-1** against

Source: Data from the American Kennel Club.

WHAT ARE THE ODDS ON BEING ONE FOR THE BOOKS?

The last detailed census turned up 121,412 librarians who can truly be said to be those who are for the books.

The Odds: **1,678-1** against

Source: Department of Commerce, Bureau of the Census, 1970.

WHAT ARE THE ODDS OF STRIKING OIL OR GAS?

In 1977, 9,961 exploratory wells were drilled. Of these, 6,101 (61.25%) were new field wildcats as opposed to added wells in known oil or gas fields. One thousand and four of these wells hit paydirt for a success rate of 16.5%.

The Odds: **5.2-1** against

Source: *Oil and Gas Journal,* "U.S. Operators Find 1,004 Fields in 1977," July 3, 1978, p. 32, 33.

WHAT ARE THE ODDS THAT IT'S ALL GREEK TO YOU?

Since only some 343,000 people speak Greek the odds are in favor of your not understanding.

The Odds: **580-1** for

Source: *Statistical Abstract,* 1979.

WHAT ARE THE ODDS THAT IT'S A LEMON?

Compared to other citrus fruits the odds are not in favor. Since they aren't sold by the dozen we counted the number of boxes produced, 26.1 million boxes at about 76 pounds net.

The Odds: **11.4-1** against

Source: Ibid.

WHAT ARE THE ODDS ON ANY FRIDAY BEING FRIDAY THE 13th?

Every 28 years there are 48 Friday the thirteenths.

The Odds: **29.4-1** against

Source: Perpetual Calendar (1800-2059)

WHAT ARE THE ODDS THAT DIAMONDS ARE A GIRL'S BEST FRIEND?

Based on women over the age of fifteen, 44% of women own diamonds. This figure doesn't include diamond engagement rings, either.

The Odds: **1.3-1** against

Source: Heron House, *The Book of Numbers*, New York: A & W Publishers, Inc., 1978.

WHAT ARE THE ODDS IT'S FOR THE BETTER?

Most people feel they are very happy or fairly happy and that five years from now will be even happier. With that kind of optimism things can only get better.

The Odds: **2.9-1** for

Source: Heron House, *The Book of Numbers*, New York: A & W Publishers, Inc., 1978.

WHAT ARE THE ODDS OF BEING A SPRING CHICKEN?

If the 4,370 million chicks hatched in 1978 were evenly spaced out over the year the spring chicken odds weren't great.

The Odds: **3.0-1** against

Source: *Statistical Abstract*, 1979.

WHAT ARE THE ODDS THAT YOU'LL TAKE THE CHANCE?

Ladies and gentlemen place your bets. Of course on each gamble the odds are different on winning, but we can tell you which games of chance most people gamble on.

The Odds: against

Legal		Illegal	
Legal	1.3-1	**Illegal**	7.9-1
Horses at track	6.3-1	Sports books	51.6-1
OTB, New York	6.4-1	Horse books	40.6-1
Legal casinos	9.6-1	Numbers	32.3-1
Bingo	4.4-1	Sports cards	32.3-1
Lotteries	3.2-1		

Source: Ibid.

WHAT ARE THE ODDS THAT THERE ARE PLENTY OF FISH IN THE OCEAN?

Absolutely splendid. According to the U.S. National Oceanic and Atmospheric Administration and the United Nations, over 73,501,000 metric tons were caught in 1977.

The Odds: **Millions-1** for

Source: Data from U.S. National Oceanic and Atmospheric Administration and the United Nations.

WHAT ARE THE ODDS OF BEING BEHIND THE EIGHT BALL?

The leading journal in the billiards field reports that an estimated 10 million different people play pool in any given year.

The Odds: **20.9-1** against

Source: Billiards *Digest*

WHAT ARE THE ODDS THAT THE ODDS ARE AGAINST IT?

We spoke to the Grand Secretary of the Grand United Order of Odd Fellows. He told us that their 18,000 members are devoted to "good works." So there are no specific items that this worthy group is against when it comes to evil.

The Odds: **18,000-0** against

Source: Data from the Odd Fellows.

WHAT ARE THE ODDS OF KEEPING A CIVIL TONGUE IN YOUR HEAD?

The population at the end of 1977 was 217,739,000. During that year 15,274,746 people were employed as full and part-time civil servants at the federal, state and local level.

The Odds: **13.2-1** against

Source: Data from Department of Commerce.

WHAT ARE THE ODDS ON A PERFECT DAY FOR A PICNIC?

Throughout these randomly selected cities we have considered only the temperature (over 65) and rain. Neither smog, smoke, haze nor the amount of rain is considered. So in real life Honolulu is a better place for a picnic than Los Angeles.

The Odds: for unless noted

	Jan	Feb	Mar	Apr	May	Jun
Austin, TX	—	—	—	1.1	1.2	2.8
Albuquerque, NM	—	—	—	—	—	2.0
Anchorage, AK	—	—	—	—	—	—
Atlanta, GA	—	—	—	—	1.2	1.1
Boston, MA	—	—	—	—	—	1.1 a
Charleston, SC	—	—	—	—	1.1	1.1 a
Charleston, WV	—	—	—	—	—	Even
Charlotte, NC	—	—	—	—	1.2	1.3
Chicago, IL	—	—	—	—	—	Even
Cleveland, OH	—	—	—	—	—	1.3 a
Denver, CO	—	—	—	—	—	1.3 a
Des Moines, IA	—	—	—	—	—	Even
Detroit, MI	—	—	—	—	—	1.1 a
Fargo, ND	—	—	—	—	—	—
Hartford, CT	—	—	—	—	—	1.1 a
Honolulu, HI	1.8 a	1.8 a	1.6 a	2.8 a	1.6 a	2.8 a
Indianapolis, IN	—	—	—	—	—	1.3
Jackson, MS	—	—	—	1.7	1.8	2.0
Little Rock, AR	—	—	—	—	1.4	1.5
Los Angeles, CA	—	—	—	—	—	6.5
Louisville, KY	—	—	—	—	—	1.3
Miami, FL	1.4	1.8	1.8	2.8	1.1	2.8 a
Milwaukee, WI	—	—	—	—	—	Even
St. Paul, MN	—	—	—	—	—	1.3 a
Nashville, TN	—	—	—	—	1.1	1.5
New York, NY	—	—	—	—	—	Even
Omaha, NE	—	—	—	—	—	—
New Orleans, LA	—	—	—	1.7	1.8	1.3
Pittsburgh, PA	—	—	—	—	—	Even
Portland, OR	—	—	—	—	—	—
Providence, RI	—	—	—	—	—	Even
Reno, NV	—	—	—	—	—	—
Richmond, VA	—	—	—	—	1.1	1.3
St. Louis, MO	—	—	—	—	—	1.5
Salt Lake City, UT	—	—	—	—	—	1.3
Seattle, WA	—	—	—	—	—	—
Sioux Falls, SD	—	—	—	—	—	1.1 a
Tulsa, OK	—	—	—	—	1.4	1.7
Wichita, KS	—	—	—	—	1.2	1.5

The Odds: for unless noted

	Jul	Aug	Sep	Oct	Nov	Dec
Austin, TX	2.9	1.8	1.5	1.7	—	—
Albuquerque, NM	1.3 a	1.2 a	1.5	—	—	—
Anchorage, AK	1.2 a	—	—	—	—	—
Atlanta, GA	1.4 a	1.1	1.7	—	—	—
Boston, MA	1.2	1.4	—	—	—	—
Charleston, SC	1.4	1.8 a	1.3	2.1	—	—
Charleston, WV	1.2 a	1.2	1.1	—	—	—
Charlotte, NC	1.1 a	1.2	2.0	—	—	—
Chicago, IL	1.2	1.8	—	—	—	—
Cleveland, OH	1.1	1.4	—	—	—	—
Denver, CO	1.8 a	1.1	—	—	—	—
Des Moines, IA	1.2	1.6	—	—	—	—
Detroit, MI	1.6	1.6	—	—	—	—
Fargo, ND	1.2	1.4	—	—	—	—
Hartford, CT	1.4	1.6	—	—	—	—
Honolulu, HI	2.4 a	2.4 a	1.7 a	1.8 a	2.8 a	2.4 a
Indianapolis, IN	1.6	1.8	1.5	—	—	—
Jackson, MS	1.1	1.4	1.5	2.9	—	—
Little Rock, AR	1.6	1.8	1.7	—	—	—
Los Angeles, CA	9.3	14.5	9.0	6.8	—	—
Louisville, KY	1.2	1.8	1.3	—	—	—
Miami, FL	2.1 a	2.4 a	2.3 a	1.6 a	1.5	2.4
Milwaukee, WI	1.1	1.6	—	—	—	—
St. Paul, MN	1.2	1.6	—	—	—	—
Nashville, TN	1.2	1.8	1.3	—	—	—
New York, NY	1.4	1.4	1.7	—	—	—
Omaha, NE	1.4	—	—	—	—	—
New Orleans, LA	1.6	1.4 a	Even	2.9	—	—
Pittsburgh, PA	1.1	1.4	—	—	—	—
Portland, OR	4.1	3.4	—	—	—	—
Providence, RI	1.2	1.2	—	—	—	—
Reno, NV	4.2	4.2	—	—	—	—
Richmond, VA	1.4	1.4	2.3	—	—	—
St. Louis, MO	1.6	1.8	Even	—	—	—
Salt Lake City, UT	2.1	1.8	—	—	—	—
Seattle, WA	2.4	2.1	—	—	—	—
Sioux Falls, SD	1.2	1.4	—	—	—	—
Tulsa, OK	2.4	2.1	1.5	—	—	—
Wichita, KS	1.8	1.8	1.5	—	—	—

All odds are -1.
a = against

Source: National Oceanic and Atmospheric Administration, *Airport Climatological Summary*, No. 90 (1965-1974), Asheville, N.C., National Climatic Center, 1979.

WHAT ARE THE ODDS ON COOKING YOUR OWN GOOSE?

In 1974 a total of 6,858 farms raised geese. Less than one in four of these sell the bird commercially so we assumed they raise them to eat and compared them to total households at the time.

The Odds: **10,191-1** against

Source: Data from Agriculture Census 1974.

WHAT ARE THE ODDS THAT SOMEONE COULD READ YOU THE RIOT ACT?

Of the population that is 14 years of age or older 1% are illiterate, and thus would not be able to read the famous statute.

The Odds: **99-1** against

Source: *Statistical Abstract,* 1979.

WHAT ARE THE ODDS THAT ANY RANDOMLY SELECTED LETTER OF THE ALPHABET WILL BE IN ORDINARY ENGLISH USAGE?

Unless of course you read the dictionary for pleasure, your basic working vocabulary is a mere 400 words. As any cryptologist would tell you, e's are the most used letter in the alphabet, but you may be surprised to find that t's beat out a's and that z deserves to be last place in the alphabet. Using any vowel is 2.7 to 1.

The Odds: against

A	11.9-1	N	13.0-1
B	77.0-1	O	13.0-1
C	32.0-1	P	55.0-1
D	23.0-1	Q	187.0-1
E	8.4-1	R	16.8-1
F	39.0-1	S	12.8-1
G	55.0-1	T	11.2-1
H	16.4-1	U	31.0-1
I	12.3-1	V	580.1
J	169.0-1	W	48.0-1
K	106.0-1	X	203.0-1
L	25.0-1	Y	50.0-1
M	33.0-1	Z	425.0-1

Source: Sports Products, Inc.

WHAT ARE THE ODDS ON DRAWING SELECTED LETTERS ON THE FIRST TURN OF SEVEN TILES IN SCRABBLE?

If you're looking for QUIZ(Blank)ES, you might as well forget it!

The Odds: against unless noted

	1 of the Letter	2 of the Letter	3 of the Letter
E	6.5-1 for	4.0-1	27.0-1
A & I	1.9-1 for	7.4-1	65.0-1
O	1.4-1 for	9.5-1	9.05-1
N,R,&T	1.3-1	17.0-1	253.0-1
D,L,S&U	2.5-1	41.0-1	1,191.0-1
G	3.6-1	81.0-1	4,683.0-1
B,C,F,H,M, P,V,W,Y, Blank	6.0-1	235.0-1	0
J,K,Q,X,Z	13.0-1	0	0

Source: Heron House Scrabble Board.

WHAT ARE THE ODDS ON IMMIGRANTS HAVING CERTAIN CHARAC- TERISTICS?

Although immigration has generally been on the decline there are certain charactristics that have predominated since 1971.

The Odds: against unless noted

Male	1.2-1
Married	3.3-1
Single	3.4-1
Female	1.1-1 for
Married	2.4-1
Single	3.7-1
Age	
Under 16 years	2.8-1
16-44 years	1.4-1 for
45 years and over	5.9-1
Occupation	
Professional, tech., and kindred	8.4-1
Managers and administrators	37.5-1
Sales workers	124.0-1
Clerical	26.0-1
Craftsmen	19.0-1

The Odds: against unless noted

Operatives (except transport)	18.2-1
Transport equipment operatives	199.0-1
Laborers (except farm)	26.0-1
Farmers and farm managers	499.0-1
Farm laborers and farm foremen	61.5-1
Service workers (except private household)	24.6-1
Private household workers	46.6-1
No occupation*	1.5-1 for

*This includes dependent women and children.

Source: *Statistical Abstract*, 1979.

WHAT ARE THE ODDS ON WHERE AN IMMIGRANT LAST LIVED BEFORE COMING TO THE UNITED STATES?

We thought you might be interested in how this has changed over the years so we're giving you two cumulations. Could it be that Australia and New Zealand are just too far away for anyone to make the trip?

The Odds: against unless noted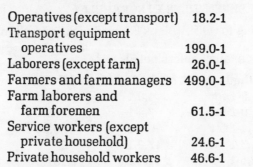

	1820-1977	1971-1977
Europe	3.1-1 for	3.9-1
Asia	17.5-1	2.1-1
Canada & South America	4.5-1	1.2-1
Africa	332.0-1	61.5-1
Australia & New Zealand	499.0-1	165.0-1
All other	165.0-1	249.0-1

Source: *Statistical Abstract*, 1979.

SELECTED BIBLIOGRAPHY

Administrative Office of the United States Courts, *Federal Offenders in U.S. District Courts 1974*, Washington, DC.

Administrative Office of the United States Courts, *Tables of Bankruptcy Statistics*, Washington, DC, 1978.

Admissions Testing Program of the College Board, *National College-Bound Seniors, 1979*.

Bell, Alan P., and Weinberg, Martin S., *Homosexualities*, New York, Simon and Schuster, 1978.

Congressional Directory 1979, 96th Congress, 1st Session, Washington, DC, Government Printing Office, 1979.

Highway Loss Data Institute, *Automobile Insurance Losses, Injury Coverages, Claim Frequency Results for 1974, 1975 and 1976 Models*, Washington, DC, 1978.

Highway Loss Data Institute, *Automobile Insurance Losses, Claim Frequency Results for 1977 Models*, Washington, DC, 1978.

Highway Loss Data Institute, *Automobile Insurance Losses, Collision Coverages, Variations by Make and Series, 1977 Models During Their First Year, 1976 Models During Their First Two Years, 1975 Models During Their First Three Years*, Washington, DC, 1977.

Hunt, Morton, *Sexual Behavior in the 1970's*, New York, Dell Publishing Co., Inc., 1974.

Insurance Information Institute, *Insurance Facts 1979*, New York, 1979.

Insurance Institute for Highway Safety, *Highway Loss Reduction Status Report*, Washington, DC, 1979.

Kinsey, Alfred, Wardell Pomeroy, Clyde Martin, *Sexual Behavior in the Human Male*, Philadelphia, W.S. Saunders Co., 1948.

National Automobile Theft Bureau, *1978 Annual Report*, Jericho, NY.

National Criminal Justice Information and Statistics Service, *Criminal Victimization in the United States 1976*, Washington, DC, Government Printing Office, 1979.

National Safety Council, *Accident Facts 1978*, Chicago, 1978.

National Safety Council, *Accident Facts 1979*, Chicago, 1979.

National Safety Council, *Fleet Accident Rates*, Chicago, 1978.

National Safety Council, *Work Injury and Illness Rates*, Chicago, 1978.

National Safety Council, *Work Injury and Illness Rates*, Chicago, 1979.

National Transportation Safety Board, Technical Information Service, *Annual Review of Aircraft Accident Data*, NTSB-ARC-76-1, Springfield, VA.

National Transportation Safety Board, Technical Information Service, *Annual Review of Aircraft Accident Data*, NTSB-ARC-78-1, Springfield, VA.

National Transportation Safety Board, Technical Information Service, *Annual Review of Aircraft Accident Data*, NTSB-ARC-78-2, Springfield, VA.

National Transportation Safety Board, Technical Information Service, *Aircraft Accident Reports*, NTSB-BA-78-8, Springfield, VA.

Neft, David S., Richard M. Cohen, and Jordan A. Deutsch, *The World Book of Odds*, New York, Grosset & Dunlap, 1978.

U.S. Department of Commerce, Bureau of the Census, *Demographic, Social and Economic Profile of States: Spring 1976*, Washington, DC, Government Printing Office, 1979.

U.S. Department of Commerce, Bureau of the Census, *Number, Timing, and Duration of Marriages and Divorces in the United States: June 1975*, P-20, No. 297, Washington, DC, Government Printing Office, 1976.

U.S. Department of Commerce, Bureau of the Census, *Population Profile of the United States: 1977*, P-20, No. 324, Washington, DC, Government Printing Office, 1978.

U.S. Department of Commerce, Bureau of the Census, *Population Profile of the United States: 1978*, P-20, No. 336, Washington, DC, Government Printing Office, 1979.

U.S. Department of Commerce, Bureau of the Census, *Statistical Abstract of the United States*, Washington, DC, Government Printing Office, 1979.

U.S. Department of Commerce, Bureau of the Census, *Voting and Registration in the Election of November 1978*, Washington, DC, Government Printing Office, 1979.

U.S. Department of Health, Education, and Welfare, Education Division, National Center for Education Statistics, *Digest of Education Statistics 1979*, Washington, DC, 1979.

U.S. Department of Health, Education, and Welfare, Public Health Service, Alcohol, Drug Abuse, and Mental Health Administration, National Institute on Drug Abuse, *Drugs and the Class of '78: Behaviors, Attitudes, and Recent National Trends*, Rockville, MD, 1979.

U.S. Department of Health, Education, and Welfare, Public Health Service, Alcohol, Drug Abuse, and Mental Health Administration, National Institute on Drug Abuse, *Highlights from Drugs and the Class of '78: Behaviors, Attitudes, and Recent National Trends*, Rockville, MD, 1979.

U.S. Department of Health, Education, and Welfare, Public Health Service, Alcohol, Drug Abuse, and Mental Health Administration, National Institute on Drug Abuse, *Highlights from the National Survey on Drug Abuse: 1977* Rockville, MD, 1979.

U.S. Department of Health, Education, and Welfare, Public Health Service, Alcohol, Drug Abuse, and Mental Health Administration, National Institute on Drug Abuse, *Marijuana and Health*, Rockville, MD, 1977.

U.S. Department of Health, Education, and Welfare, Public Health Service, Alcohol, Drug Abuse, and Mental Health Administration, National Institute on Drug Research, *National Survey on Drug Abuse: 1977, Volume I, Main Findings*, (ADM) 78-618, Rockville, MD, 1977.

U.S. Department of Health, Education, and Welfare, Public Health Service, Center for Disease Control, *Abortion Surveillance 1976*, (CDC) 78-8205, Atlanta, Georgia, 1978.

U.S. Department of Health, Education, and Welfare, Public Health Service, Health Resources Administration, National Center for Health Statistics, *Dietary Intake Findings United States, 1971-1974*, Hyattsville, MD, 1977.

U.S. Department of Health, Education, and Welfare, Public Health Service, Health Resources Administration, National Center for Health Statistics, *Food Consumption Profiles of White and Black Persons Aged 1-74 Years: U.S. 1971-74*, Hyattsville, MD, 1979.

U.S. Department of Health, Education, and Welfare, Public Health Service, National Institute on Alcohol Abuse and Alcoholism, *Alcohol and Health, New Knowledge*, Rockville, MD, 1974.

U.S. Department of Health, Education, and Welfare, Public Health Service, National Institute on Alcohol Abuse and Alcoholism, *Third Special Report to the U.S. Congress on Alcohol and Health*, Washington, DC, Government Printing Office, 1978.

U.S. Department of Health, Education, and Welfare, Public Health Service, Office of Health Research, Statistics, and Technology, National Center for Health Statistics, *Plan and Operation of the Health and Nutrition Examination Survey, United States 1971-1973*, Series 1, No. 10a, Hyattsville, MD, Government Printing Office, 1973.

U.S. Department of Health, Education, and Welfare, Public Health Service, Office of Health Research, Statistics, and Technology, National Center for Health Statistics, *Plan and Operation of the Health and Nutrition Examination Survey, United States 1971-1973*, Series 1, No. 10b, Hyattsville, MD, Government Printing Office, 1973.

U.S. Department of Justice, *F.B.I. Uniform Crime Reports, Crime in the United States 1978*, Washington DC, Government Printing Office, 1978.

U.S. Department of Justice, Law Enforcement Assistance Administration, National Criminal Justice Information and Statistics Service, *Crime Against Persons*, Washington DC, Government Printing Office, 1979.

U.S. Department of Justice, Law Enforcement Assistance Administration, National Criminal Justice Information and Statistics Service, *Criminal Victimization in the United States 1976*, Washington DC, Government Printing Office, 1976.

U.S. Department of Justice, Law Enforcement Assistance Administration, National Criminal Justice Information and Statistics Service, *Rape Victimization in 26 American Cities*, Washington DC, Government Printing Office, 1979.

U.S. Department of Justice, Law Enforcement Assistance Administration, National Criminal Justice Information and Statistics Service, *Sourcebook of Criminal Justice Statistics—1978*, Washington DC, Government Printing Office, 1978.

U.S. Department of Labor, Bureau of Labor Statistics, *U.S. Working Women: A Databook*, Washington, DC, Government Printing Office, 1977.

U.S. Department of Transportation, National Highway Traffic Safety Administration, *Fatal Accident Reporting System, 1977 Annual Report*, Washington, DC, Government Printing Office.

U.S. Department of Transportation, National Highway Traffic Safety Administration, *Fatal Accident Reporting System, 1977 Annual Report, Volume II*, Washington, DC, Government Printing Office.

U.S. Department of Transportation, National Highway Traffic Safety Administration, *Fatal Accident Reporting System, 1978 Annual Report*, Washington, DC, Government Printing Office.

U.S. Department of Transportation, Research and Special Programs Administration, Office of Policy, Plans and Administration, Transportation Systems Center, Transportation Information Division, *National Transportation Statistics*, (DOT) TSC-RSPA-79-19, Washington, DC, Government Printing Office, 1979.